MW00964174

The Deep Green Planet

From The Water To The Land

The Deep Green Planet

FROM THE WATER TO THE LAND

RENATO MASSA
WITH
MONICA CARABELLA AND LORENZO FORNASARI

ENGLISH TRANSLATION BY NEIL FRAZER DAVENPORT

Austin, Texas

Published by Raintree Steck-Vaughn Publishers, an imprint of Steck-Vaughn Company

Editors
Caterina Longanesi, Linda Zierdt-Warshaw

Design and layout
Jaca Book Design Office

Library of Congress Cataloging-in-Publication Data

Massa, Renato.
[Dall'acqua alla terra. English]
From the water to the land / Renato Massa with Monica Carabella and Lorenzo Fornasari.
p. cm. — (The Deep green planet)
Includes bibliographical references and index.
Summary: Examines the evolution of plant species from their beginnings as marine plants through their colonization of dry land.
ISBN 0-8172-4310-0
1. Plants — Evolution — Juvenile literature. 2. Botany — Juvenile literature. [1. Plants — Evolution. 2. Botany.] I. Carabella, Monica. II. Fornasari, Lorenzo, 1960– . III. Title. IV. Series.
QK980.M3713 1997
581 — dc20 96-2880
CIP AC

Printed and bound in the United States
1 2 3 4 5 6 7 8 9 0 WO 99 98 97 96

Contents

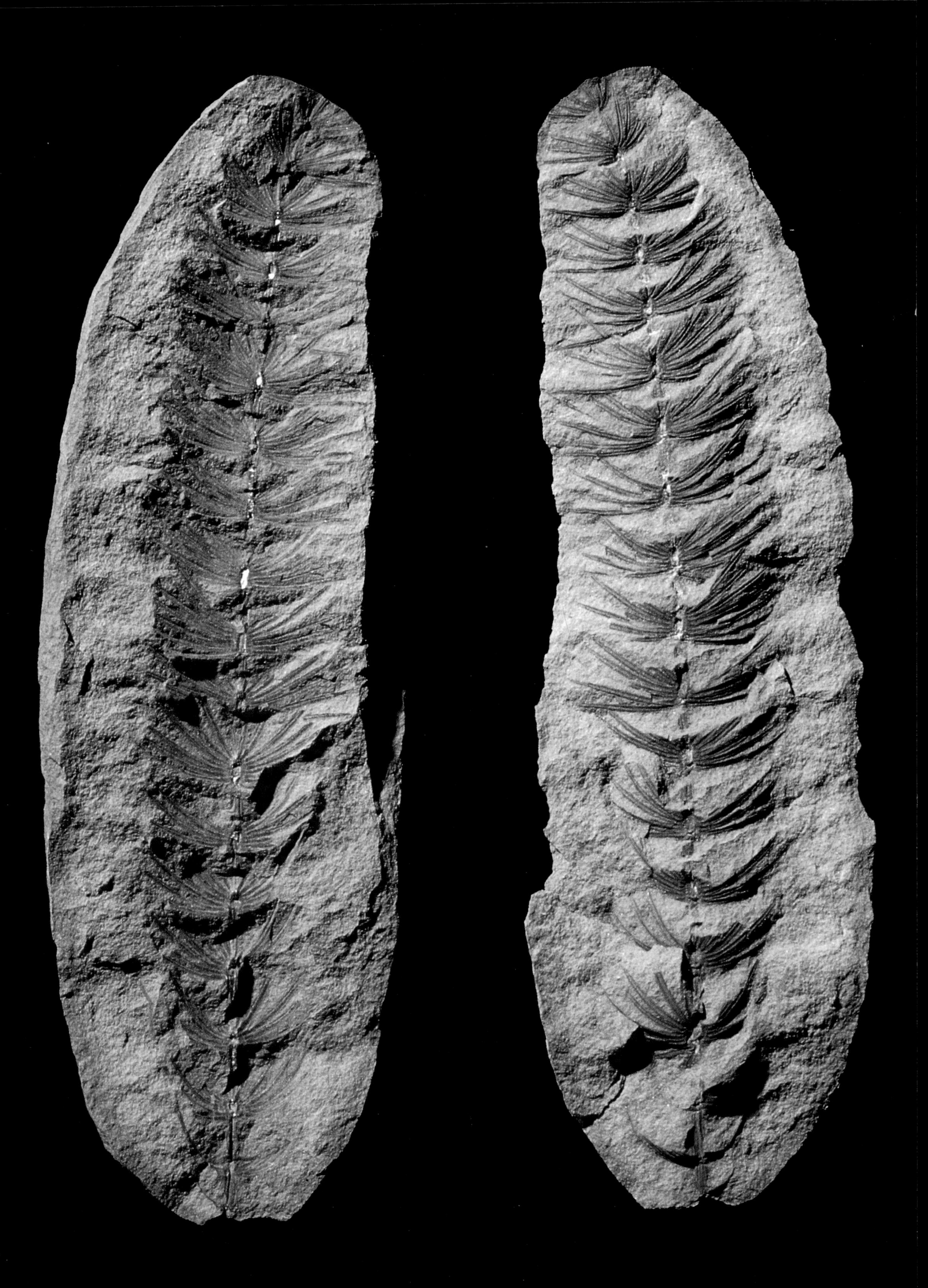

INTRODUCTION

When people think about evolution, they often imagine many pulsing cells in water. At first, the cells are separate. Then, they gather into groups. Later, the cells develop into jellyfish, planaria, shellfish, amphibians, reptiles, birds, and mammals. People rarely imagine the other direction taken by those pulsing cells—the parallel evolution that led to the creation of the plant kingdom.

Plant species evolved in the same way as animal species. The first to develop were marine plants. Later, new forms, equipped with a vascular system, evolved and colonized dry land. The passage of the plants from the water to the land was vital. Without it, animals would never have been able to populate the immense deserts of rock, sand, dust, and mud that formed the continents before they were colonized by plants.

The fundamental energy supply of plants—the use of solar energy rather than energy produced by other organisms—affected the way they evolved. Plants were also faced with the problem of surviving, flourishing, and reproducing while staying in the same place. They had to find a way to anchor themselves firmly, to transport water and mineral salts, and to reproduce out of water. Plants cannot move. Therefore, they cannot use the same systems of contact and union that animals use to bring together the male and female gametes. Plants developed alternative methods of fertilization. First, they relied on rain and dew. Later, they developed a new organ for reproduction in air—the flower. This fascinating story, which ranges from the ocean depths to the mountain uplands, is the subject of the second volume in the Deep Green Planet series.

RENATO MASSA

FROM THE WATER TO THE LAND

Along the Coast

The term **symbiosis** is used to describe a relationship between two species that live in association with each other, providing benefits to both **organisms**. The evolution of plants began as a symbiotic relationship between amoebas and bacteria. The amoebas could not carry out **photosynthesis**, but the bacteria could. Together, the amoebas and bacteria formed plantlike cells that had **chloroplasts**. The cells united with others in colonies that were probably surrounded by a sticky, soupy material given off by the cells.

Millions of years after these plantlike cells formed, many-celled plants appeared in the oceans. These plants moved onto the land along the coastlines. Today, plantlike organisms that live along the coasts are classified into four main groups. These groups include green **algae** (**chlorophytes**), brown algae (**phaeophytes**), red algae (**rhodophytes**), and yellow-brown algae (**xanthophytes**). Because the cell structure of green algae is similar to that of green plants, it is thought that all land plants developed from green algae.

Adaptation to coastal environments is advantageous to the survival of algae. The shallow and sheltered waters along the coast are suitable for photosynthesis. In addition, there is a constant supply of the mineral salts needed for plant growth. Often, the algae have flat or crustlike forms that increase their surface area for photosynthesis. Other algae have leaflike structures called **fronds**. The fronds grow upward to use the light and the movement of the water most effectively.

Living on the coastline also has its problems. Layers of sand and mud on the seabed are constantly disturbed by the action of waves. Algae must therefore anchor themselves to solid supports to keep from being washed out to sea. Since the only secure supports are rocks, algae have adapted to life in rocky coastal environments.

Toward Dry Land

Algae have developed many ways to protect themselves from the force of waves. To avoid breaking, they form strong bonds between their different cells. To move **nutrients** from their upper parts to their lower parts and to the **holdfast**, or anchor, they increase the density of cells. They also produce a flexible inner framework of **permeable** cell walls.

Algae grow mainly from their tips. Photosynthesis is most efficient at these tips. As algae grow, they use their inner cells to transport the products of photosynthesis toward their base. At the same time, their holdfasts grow larger, and true rootlike structures develop. These adaptations were forerunners of the specialized parts of higher plants, such as **roots**, stems, and **leaves**. Other adaptations demonstrate tissue specialization. For example, some algae, such as *Lessonia*, have rigid trunks with fronds that hang loosely at low tide.

These evolutionary advances prepared coastal algae for life out of the water. However, before algae could move onto dry land, they had to develop ways to avoid drying out. One way to avoid drying included waterproofing the stem through the formation of a **cuticle**. Other changes included the opening of **stomata** for the exchange of gases, the movement of the photosynthetic cells toward the inner parts of the organism, and the formation of true veins for the transport of water and mineral salts. In addition, the holdfast of algae evolved into true roots that protected against the wearing or grinding of the soil.

When ancient bacteria and blue-green bacteria capable of photosynthesis entered into symbiosis with the primitive amoebas, chloroplasts and plantlike cells with a defined nucleus developed. The first algae developed and rapidly colonized seabeds. **1.** Algae are green, red, yellow-brown, or brown. **2.** *Fucus* is a common brown alga. **3.** A green alga. **4.** Two rootlike structures of the brown alga *Laminaria*, or sea tangle, attached to the rock on which the alga was anchored. **5.** Another green alga. **6.** A red alga on a rock.

2
3
4
5
6

Fossil remains have shown that one of the first plants to have true tissues is *Rhynia*. (An illustration of *Rhynia* is shown on page 28.) *Rhynia* lived in swamps during the Devonian period, 345 to 395 million years ago. This primitive land plant had a very simple structure. It also had an underground stem called a **rhizome**. The rhizome was used to absorb water and mineral salts. At the base of the rhizome were short roots. The body of *Rhynia* had structures that branched off two by two and carried the reproductive structures of the plant.

Asteroxylon was a slightly more complex early land plant than *Rhynia*. (Compare the appearances of the two plants in the illustration on page 28.) The part of *Asteroxylon* that grew aboveground was covered by small scales. The scales were similar to leaves. However, the scales lacked the veins present in leaves that carried water and dissolved materials.

Yellow-brown algae
Xanthophytes

Brown Algae
Phaeophytes

The arrow between the two diagrams shows that only green algae succeeded in evolving and moving onto dry land. Here, the algae developed into a large variety of land plants. The lower diagram shows eight species of algae belonging to the four different groups.
1. *Botrydium granulatum*, a yellow-brown alga. The alga looks like a small button with branching rootlike structures at the bottom. **2.** *Tribonema vulgare*, a yellow-brown alga with nonbranching filaments.
3. *Padina* genus, a brown alga with fanlike sections attached to a strong, central base.
4. *Laminaria saccharina*, a brown alga with large, long, oval fronds. **5.** *Enteromorpha compressa* and *Acetabularia mediterranea* (**6**), two very different green algae.
7. *Delesseria sanguinea*, a red alga with veined leaves surprisingly similar to those of higher plants. **8.** *Plocamium coccineum*, a red alga with tiny, elegant branches.

THE REPRODUCTION OF LIFE

Copying or Exchange

Reproductive cycles are very important to the evolution of plants. During reproductive cycles, the number of **chromosomes** in the **nucleus** of the cell changes. Either one set (n) or two sets ($2n$) of chromosomes are present at different stages of the reproductive cycle. Often, there is also an **alternation of generations**. An alternation of generations is a life cycle in which the organism exists in different forms at different stages of the cycle.

The simplest single-celled algae do not reproduce sexually. Single-celled algae multiply by cell division. In cell division, an organism splits, forming two individual cells with the same genetic material. This type of cell division is called **binary fission**.

The result of binary fission is the same as that of **mitosis.** Mitosis is cell reproduction in which the nucleus makes an exact copy of itself. During mitosis, no changes are made to the genetic material of the cell.

Binary fission and mitosis are processes in which **genes** are copied. The processes have nothing to do with sexuality, in which a transfer or exchange of genetic material occurs. Sexuality first appeared in bacteria in the process of **conjugation.** Conjugation involves an exchange of genetic material without reproduction.

Sexuality and reproduction combine in the process of **meiosis**. During meiosis, a cell with two sets of chromosomes, called a $2n$ or a **diploid** cell, divides to form four new cells. Each new cell, called an n or a **haploid** cell, has a single set of chromosomes. These single sets of chromosomes all differ from the original chromosome. The difference is caused by an exchange of genetic material between the pairs of chromosomes in the original cell. During **sexual reproduction**, two haploid cells fuse. In this process, they combine their chromosomes to again form a diploid cell.

Reproductive Cycles

There are several kinds of reproductive cycles. Sexual reproduction usually involves specialized cells called **gametes**. A **zygote** is a diploid cell produced by the union of a male gamete, called **spermatozoa**, and a female gamete, called an **egg**. The **haplont cycle** leads to the production of a zygote. The zygote immediately divides by meiosis to produce four cells, each having a single set of chromosomes. These cells are called **spores**. Spores are cells that can develop into many-celled adult organisms without joining with other cells. The adult organisms then develop new gametes through mitosis.

In the **diploid cycle**, sexual reproduction gives rise to a zygote. Through mitosis, the zygote produces a many-celled organism, usually an animal. This organism forms gametes through meiosis. The gamete produced by another organism must join with a gamete produced by an organism of the same species, but opposite gender, to form a new organism.

During the **diplohaplont cycle**, the zygote produced through sexual reproduction gives rise to a new organism through meiosis. The new organism is called a **sporophyte**. The sporophyte makes spores that develop into organisms called **gametophytes**. Through meiosis, gametophytes produce gametes that

2

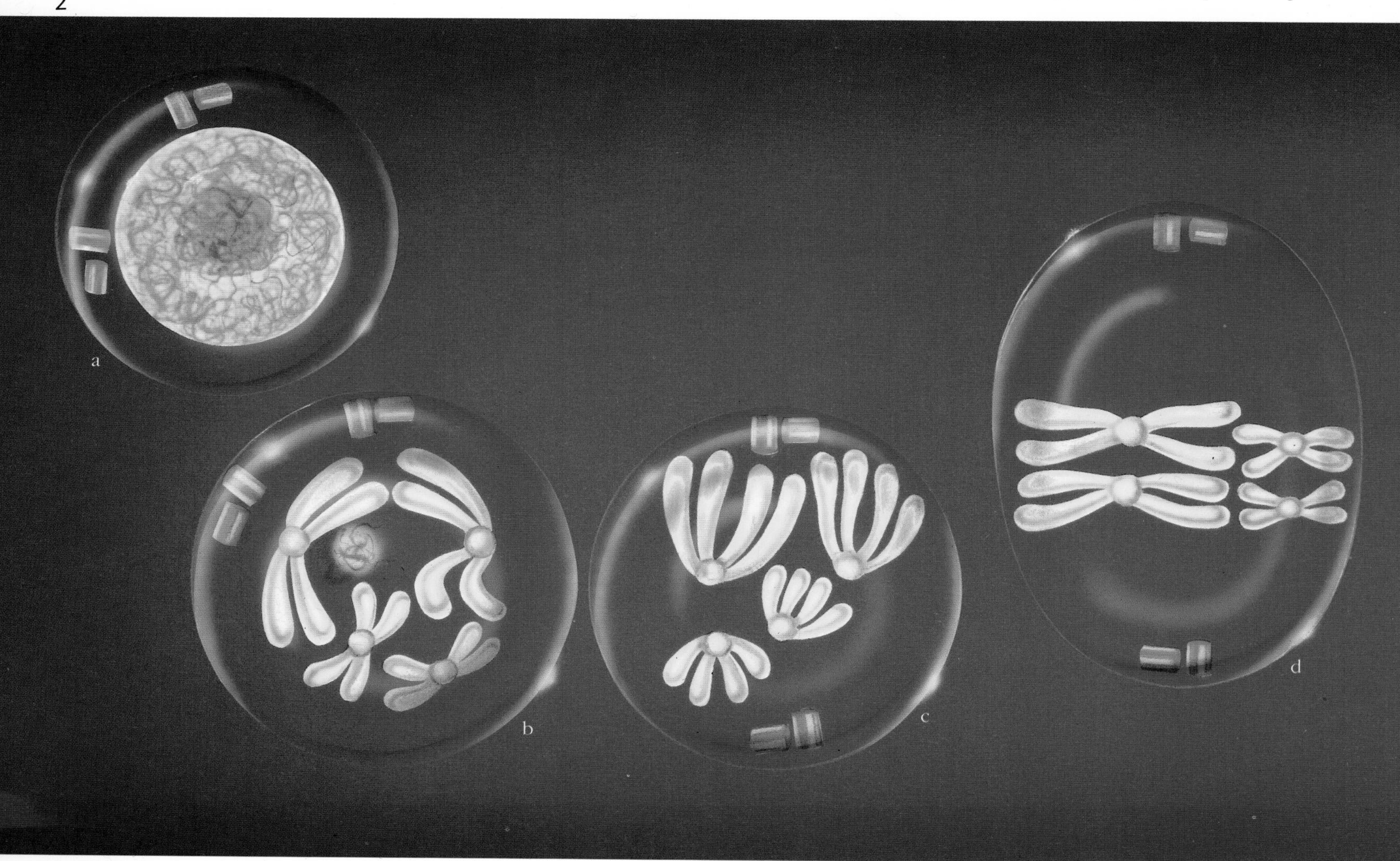

The diagrams show mitosis in a eukaryote cell (**2**) and binary fission in a bacterium (**1**). In both mitosis and binary fission, the cell divides into two after producing a copy of its genetic material. The more simple process is that of binary fission. **1.** In the bacterium, all the genetic material is contained in a single circular chromosome (**a**). After the genetic material has been duplicated, the cell begins to pinch at the middle (**b**, **c**, and **d**) and divides into two daughter cells (**e** and **f**). Each daughter cell has a circular chromosome formed by the duplication of the genetic material. **2.** In the case of mitosis, the process is more complicated, but the result is the same. **a.** The nucleus contains bundled genetic material. Small, cylindrical bundles outside the nucleus become visible. **b.** The bundled genetic material has organized into chromosomes. **c.** The cylindrical bundles migrate to opposite poles of the cell. **d.** The chromosomes arrange themselves at the center of the cell, which begins to lengthen. **e.** The chromosomes separate into two identical halves and move toward the cylindrical bundles. **f.** The mother cell pinches at the middle, leaving a pair of cylindrical bundles and a complete set of chromosomes in each half. **g.** Two separate and identical cells form.

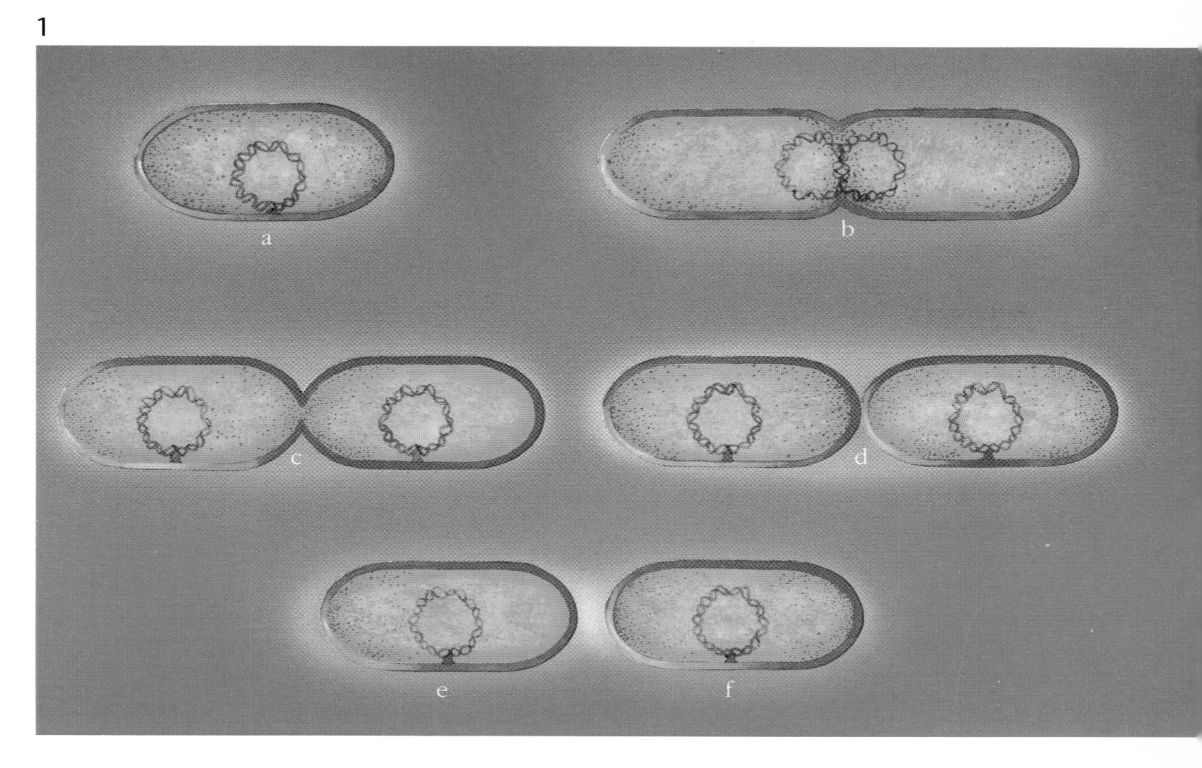

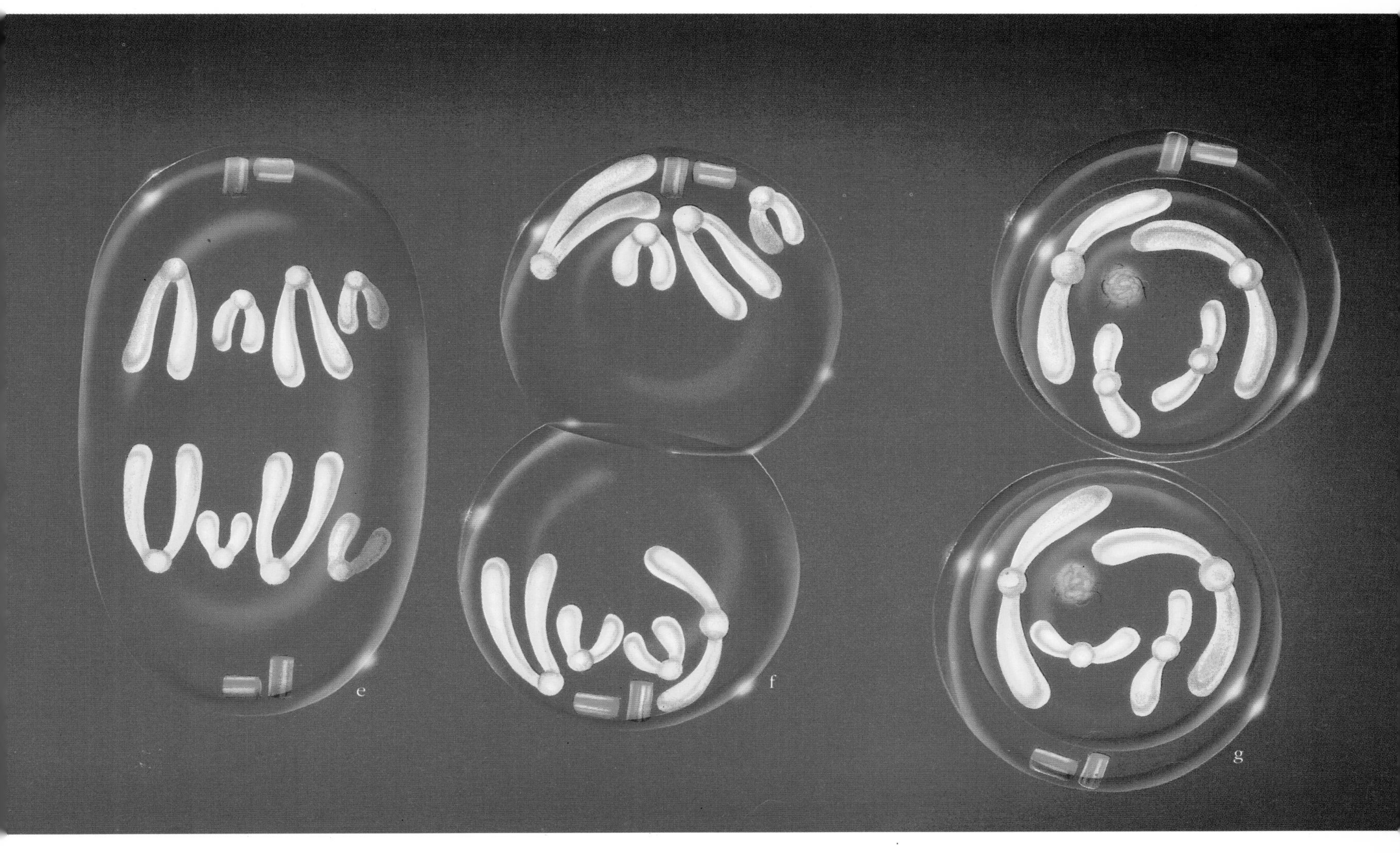

each contain a single set of chromosomes. As in the haplont cycle, the gametes can unite to produce a zygote.

Alternation of Generations

At different stages of the life cycle, a plant may exist as a sporophyte or a gametophyte, or as two forms of the same plant. The change in form of gametophytes and sporophytes is called an alternation of generations. During different stages of the alternation of generations cycle, the plant exists in a haploid or a diploid state. The change in the number of chromosomes from n to $2n$ and back again is called an **alternation of nuclear phase**.

Many green algae do not have an alternation of generations. In such algae, the zygote formed by the union of the gametes is the only diploid cell. In some haploid species, such as *Ulothrix zonata*, meiosis forms two types of gametes that represent the two genders, male and female. Other species, such as those of the genus *Cladophora*, have an alternation of generations known as **isomorphic alternation**. In isomorphic alternation, the two generations look the same but have different numbers of chromosomes ($2n$ or n). Still other species have a **heteromorphic alternation** of generations. In a heteromorphic alternation of generations, the haploid and diploid generations differ in form.

Plant Life Cycles

Algae and plants that have a diploid cycle like those of animals are rare. Only a few green algae, **diatoms**, and some **fungi** called **oomycetes** have this kind of life cycle.

When plants left the ocean, changes in the diplohaplont cycle occurred. There was now a tendency for either the gametophyte (n) or

1

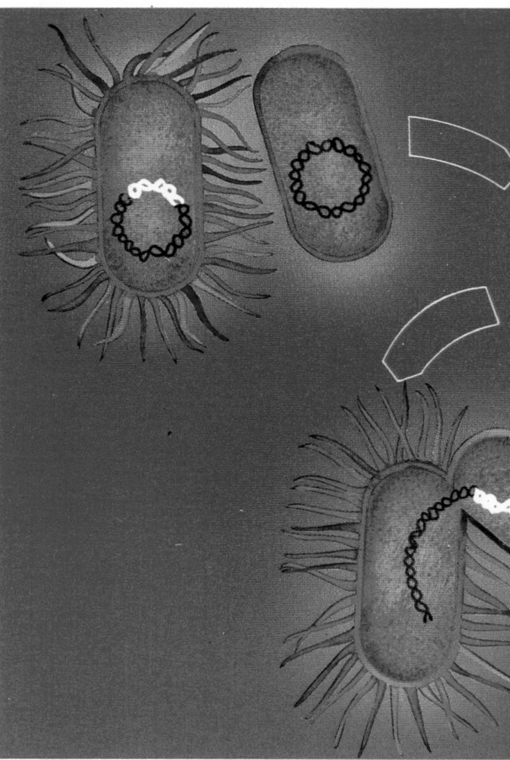

2

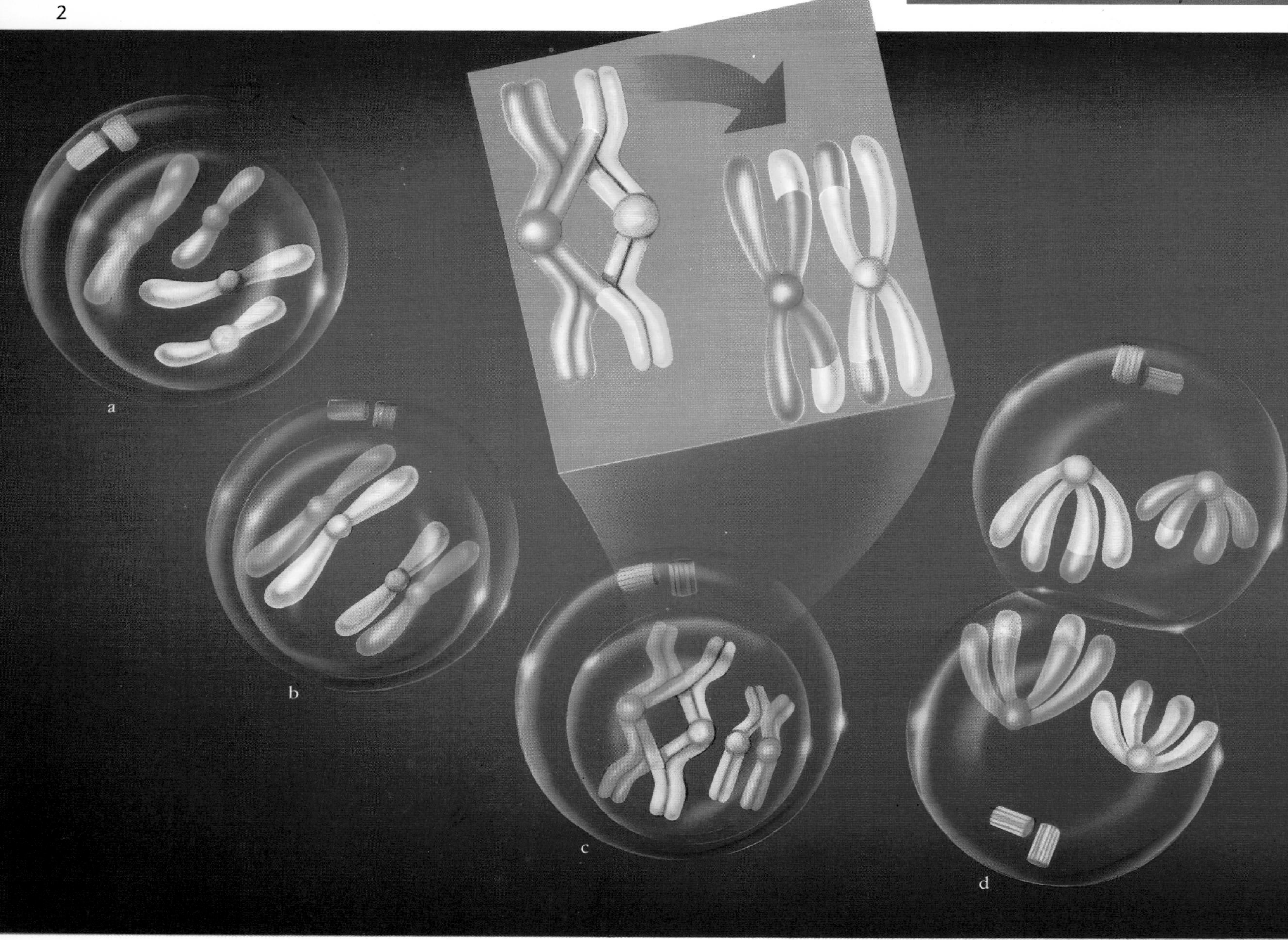

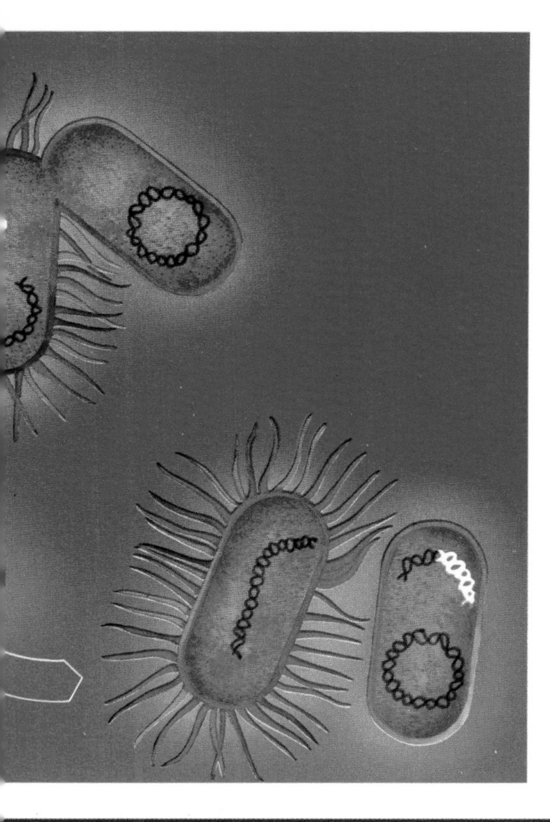

The two diagrams on these pages compare the basic processes of sexual reproduction in organisms composed of cells containing visible nuclei and single-celled organisms without visible nuclei. In both cases, there is an exchange of genes and a change in the genetic material of the organism involved.

1. The most simple sexual reproduction process is that of conjugation between single-celled organisms without visible nuclei. Two bacteria of different "sexes" come into contact using a cell organ that creates a cytoplasmic bridge on the "male." The "male" donates a segment of its genetic material to the bacterium without the organ, the "female." If the donor bacterium has a different genetic makeup, the receiver will be given a variation of some of its genes. In this way it will become partially diploid.

2. The process of meiosis is much more complicated than conjugation. Meiosis involves the germ cells of organisms containing visible nuclei, and results in the production of gametes (eggs and spermatozoa). The gametes cannot develop unless they unite or are spores that can develop on their own. **a.** The identical chromosomes of a diploid cell pair up (**b**), unite, and exchange material in the crossing over process shown in the pink square (**c**). After this exchange, the first meiotic division occurs in which (**d**) two new cells form. **e.** Each new cell has a single pair of chromosomes. **f.** At this point, each daughter cell divides. The chromosomes are pulled apart, and each of the four new cells receives half of the genetic material. **g.** The end result of the two divisions is the production of four cells. All of the cells are different from one another because of crossing over and the random assortment of the original chromosomes.

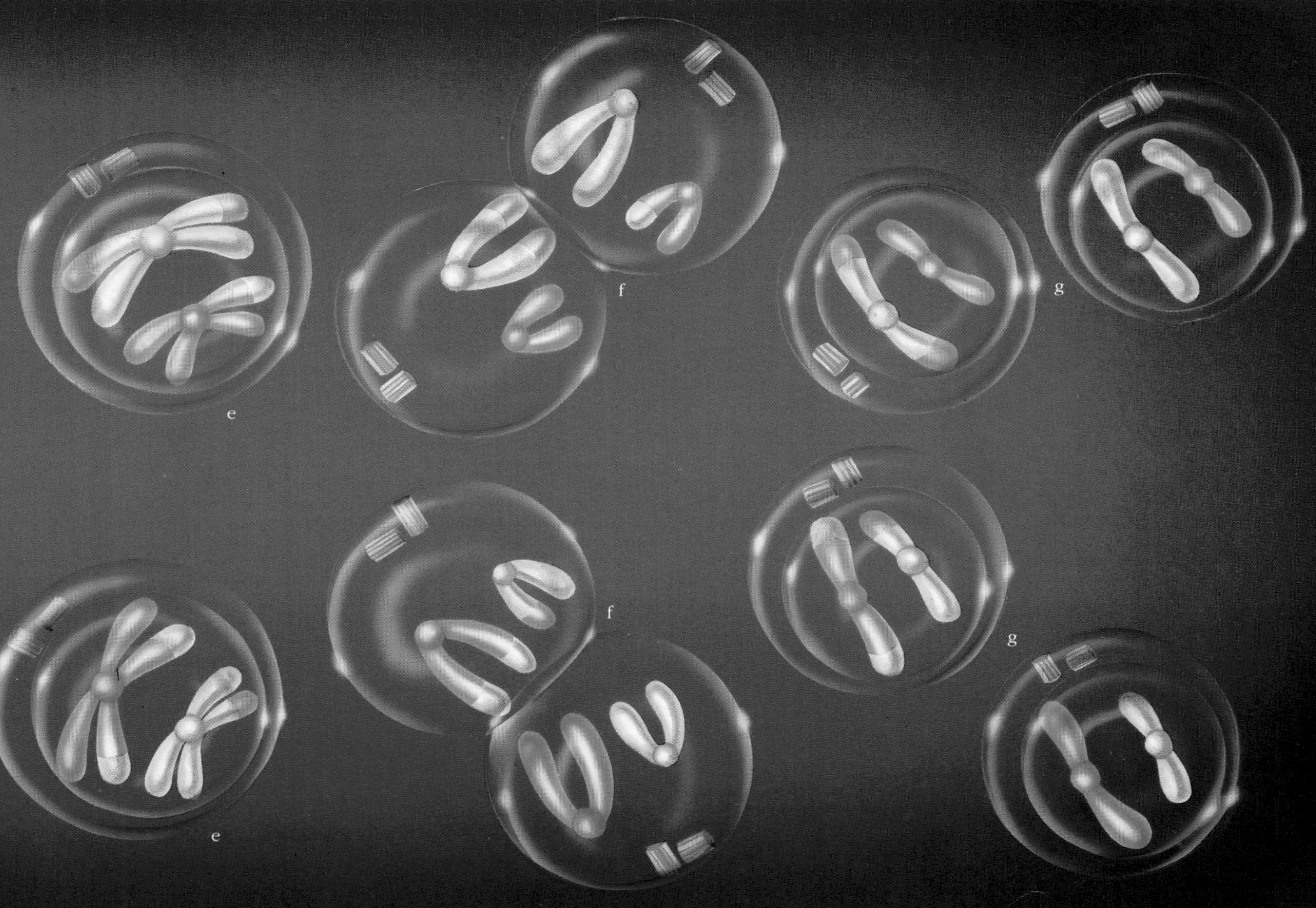

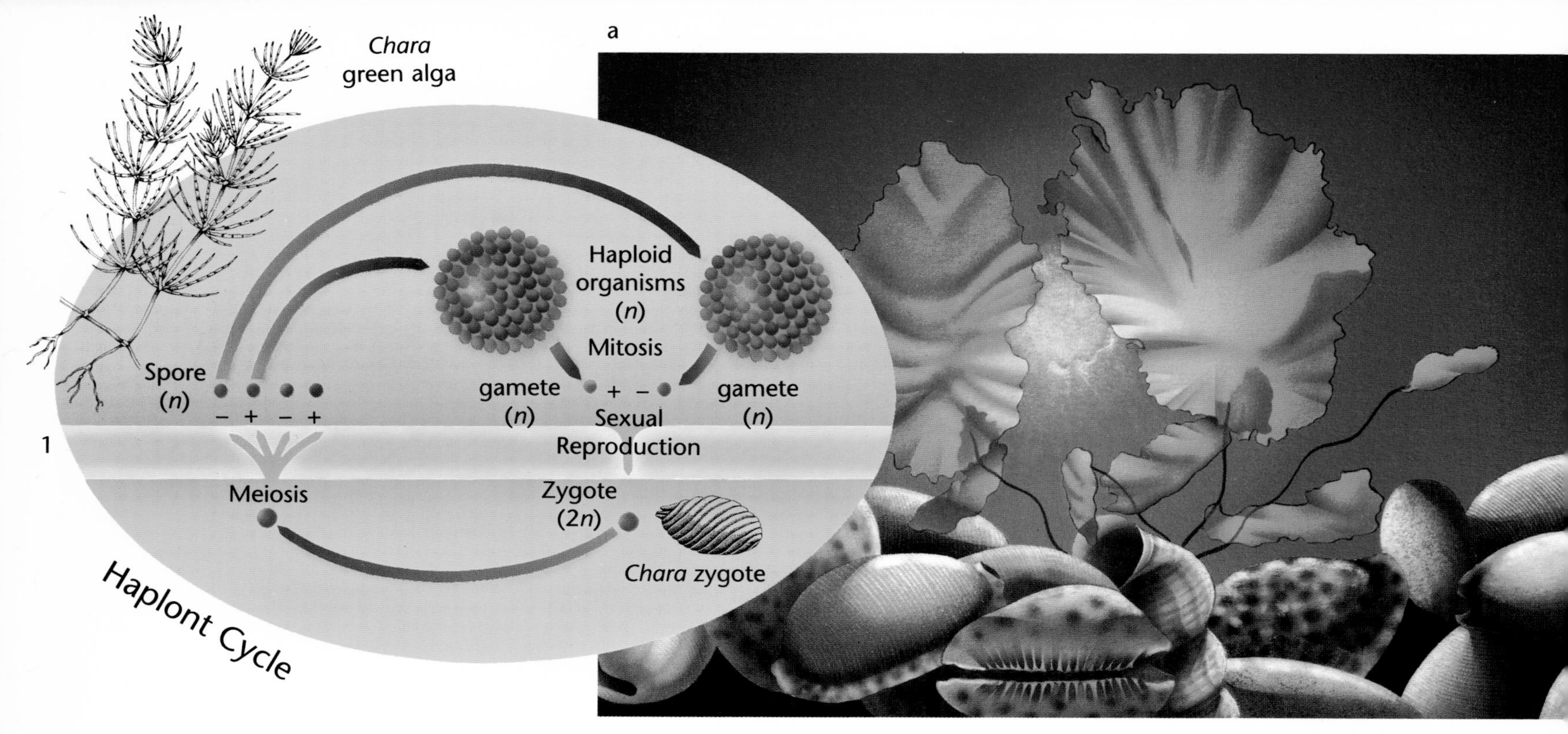

the sporophyte ($2n$) generation to dominate. The changes affected the **bryophytes** more than the highly evolved **pteridophytes** and **spermatophytes.** In the bryophytes, the gametophyte is independent of and generally larger than the sporophyte. In the pteridophytes and spermatophytes, the gametophyte is smaller. In flowering plants, or **angiosperms**, the sporophyte is almost always dominant. In algae that have a heteromorphic alternation of generations, the sporophyte is usually dominant.

1. In the haplont cycle there is an alternation of the nuclear phase from n to $2n$ and vice versa, but there are only n generations. Meiosis occurs immediately after the two gametes have united into a single zygote. The only cells capable of reproducing through mitosis are the n cells produced through meiosis. Only these cells can give rise to multicelled organisms. **a.** This cycle can be observed in the red alga *Porphyra leucostica*, in other algae, and in some fungi. **2.** In the diploid cycle there is also an alternation of the nuclear phase, but only $2n$ generations. Meiosis gives rise to gametes, or eggs and spermatozoa, which cannot develop unless they unite to form a $2n$ zygote. The zygote will undergo mitosis and give rise to a $2n$ generation. Some of the cells of this generation will undergo meiosis and produce n gametes. The diploid cycle is typical of animals, including humans represented here by a city (**b**), the protists, and some algae such as *Fucus*. **3.** In the diplohaplont cycle there is both the n generation (gametophyte) and the $2n$ (sporophyte) generation. The sporophyte is the plant we normally see. This cycle is typical of most plants from the brown alga *Laminaria* (**c**) to the chestnut *Castanea sativa* (**d**).

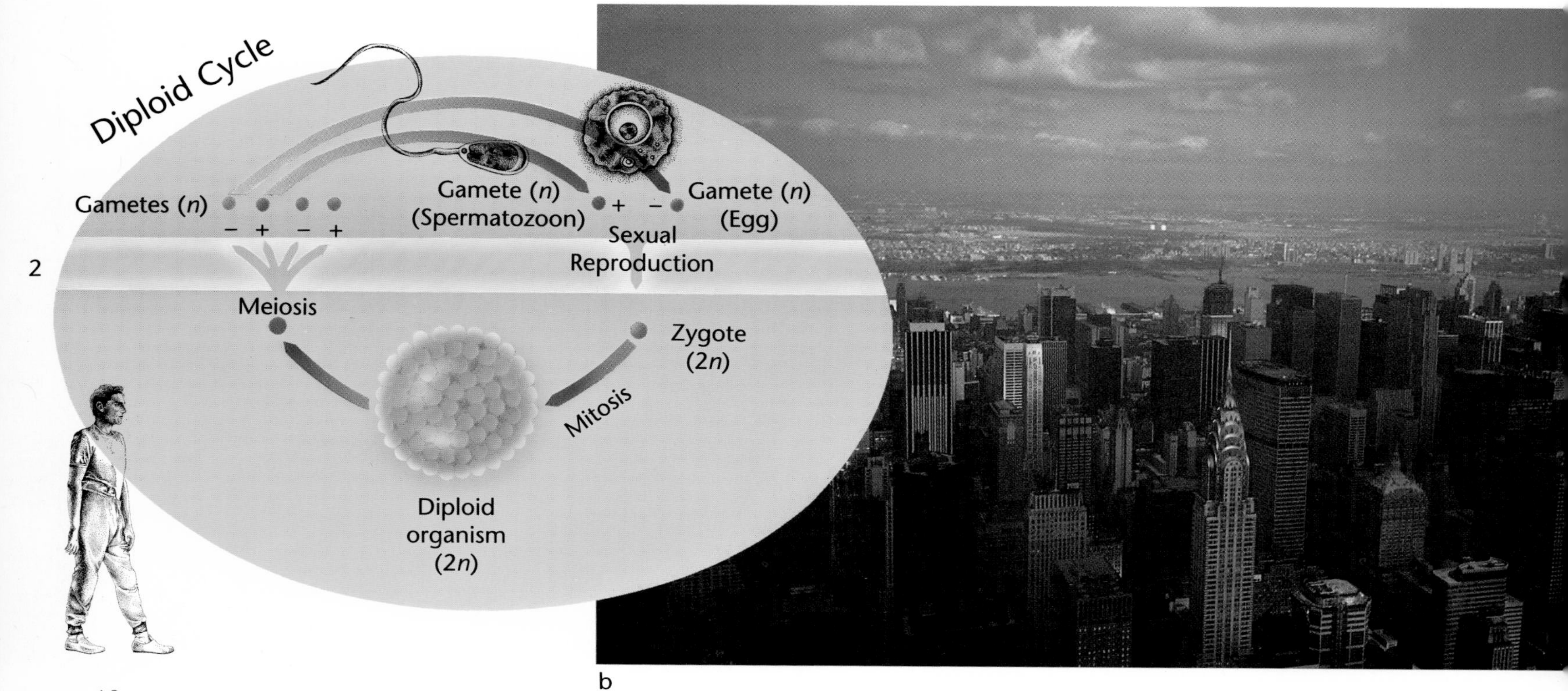

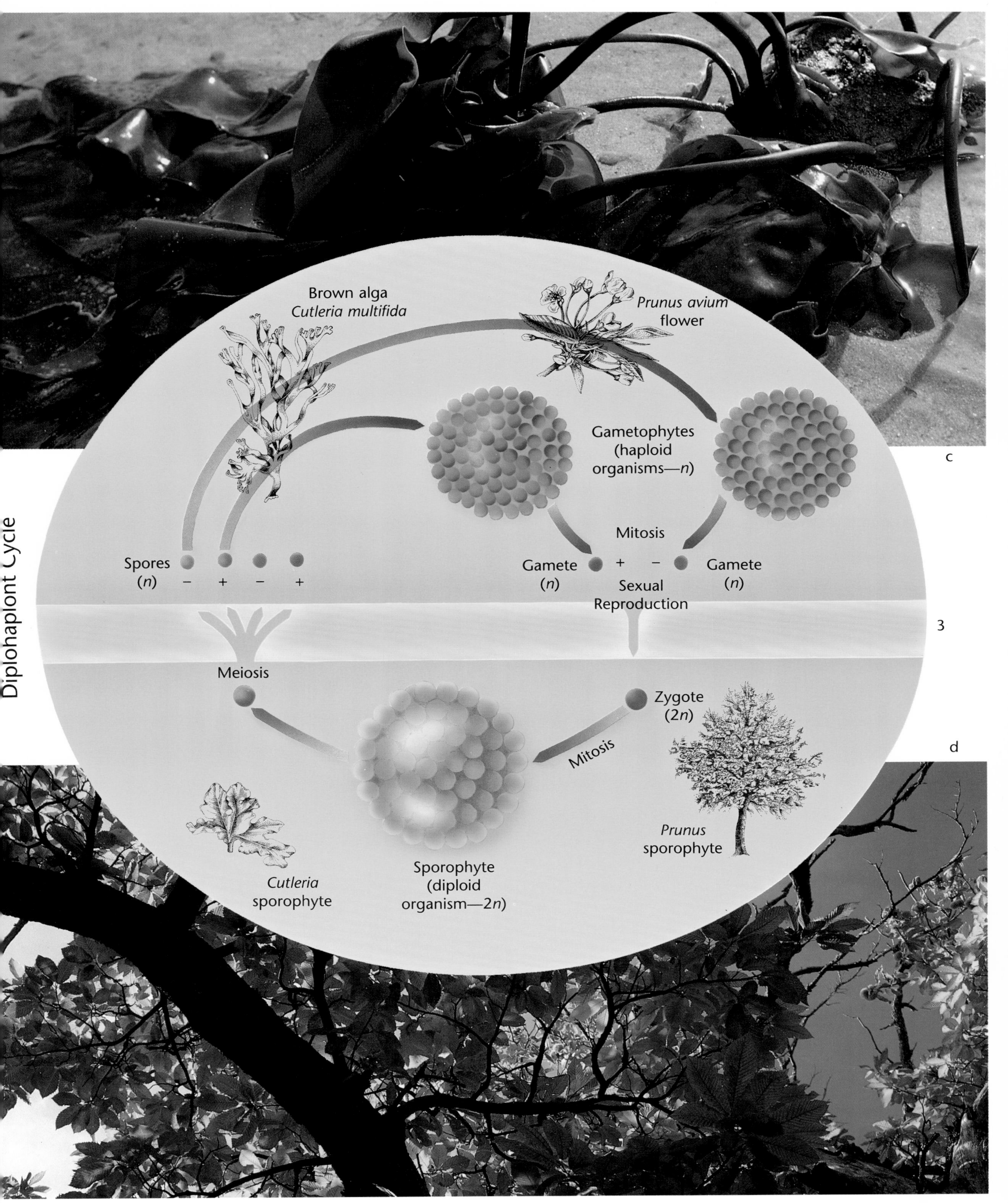
Brown alga
Cutleria multifida
Prunus avium
flower
Gametophytes
(haploid
organisms—n)
Mitosis
Spores
(n)
−
+
−
+
Gamete
(n)
+
−
Gamete
(n)
Sexual
Reproduction
Meiosis
Zygote
(2n)
Mitosis
Prunus
sporophyte
Cutleria
sporophyte
Sporophyte
(diploid
organism—2n)
c
3
d
Diplohaplont Cycle

MOSSES, LIVERWORTS, AND HORNWORTS

Ancient Land Plants

The bryophyte group, which includes mosses, **liverworts**, and hornworts, evolved alongside other plants. They were the first land plants to have the general appearance of plants, even though they did not have true plant tissues. Bryophytes have a small **rhizoid** structure that acts as a root. The rhizoid absorbs water and anchors the plant in the ground. Bryophytes also have a stem, which is either erect or creeping. The stem, which is more developed in the mosses, acts as a support for the small leaflike structures of bryophytes and the structures that produce the gametes. The plants we recognize as mosses are haploid gametophytes (n). The diploid sporophyte ($2n$) is attached to the gametophyte. The sporophyte is never independent and is fed by the gametophyte. The sporophyte looks a little like a small flower bud in the moss.

The Bryophyte Cycle

The bryophyte cycle begins with the **germination** of a spore. A spore is a reproductive cell that can develop into an adult without fusing with other cells. Spores that are produced through meiosis are called **meiospores.** When the spore germinates, it forms a thin, branching thread called the **protonema.** A number of buds grow from the protonema. Each bud gives rise to a gametophyte. The gametophyte may live for one or many years. Bog mosses, for example, grow continually to form huge carpets and may be hundreds of years old.

Male or female gametes develop inside the **gametangia.** The gametophytes carry either male gametangia, called **antheridia**, or female gametangia, called **archegonia**. Water, as rain or dew, must be present for male gametes to leave their gametangia and reach the female gametes. Fertilization of the female produces a zygote that immediately begins to divide and grow. In liverworts, the growth phase is protected by the archegonium. However, in mosses, the archegonium immediately breaks apart.

1. A carpet of *Cratoneurum* moss. Humid, shady forest undergrowth is the ideal environment for many mosses. Mosses colonize the ground and the rocks to form large green velvet cushions. **2.** A close-up of *Dicranum* moss, which looks like a sea of bright green grass. **3.** A detailed drawing of a *Sphagnum* moss plant. This moss can hold large amounts of water and may cover entire mountain slopes. The spore capsules can clearly be seen at the top. **4.** A cross section of the stem of a moss, *Mnium orthorynchum,* as seen with a scanning electron microscope. You can see what may be considered primitive vascular tissues for the transport of nutrients. **5.** A *Brachythecium* moss capsule in three stages of development. The opening at the top has a double ring of teeth. As long as the environment is humid, the teeth of the outer ring remain closed. As the environment becomes drier, the capsule dries out, and the teeth open, allowing the spores to be spread by the wind.

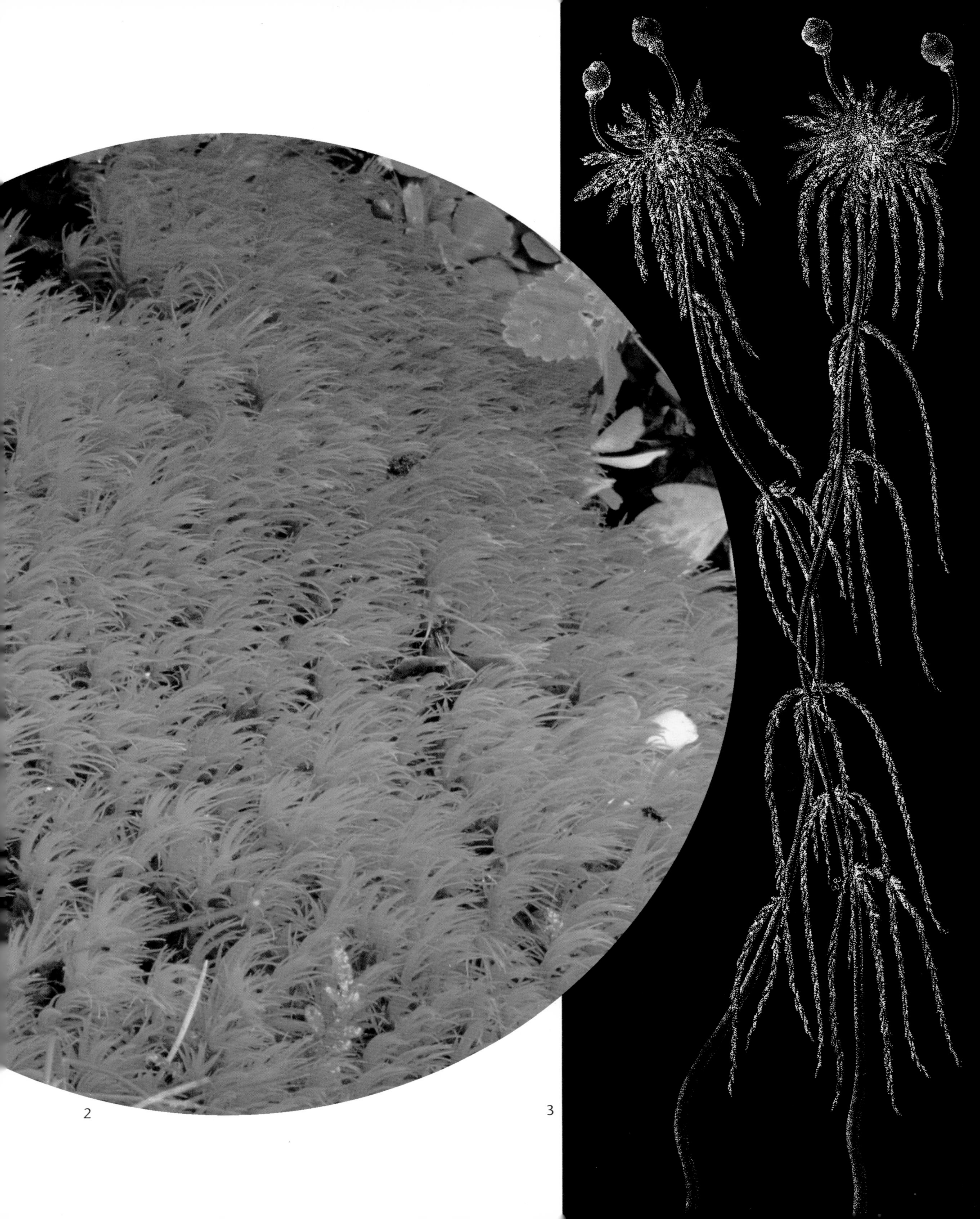
2
3

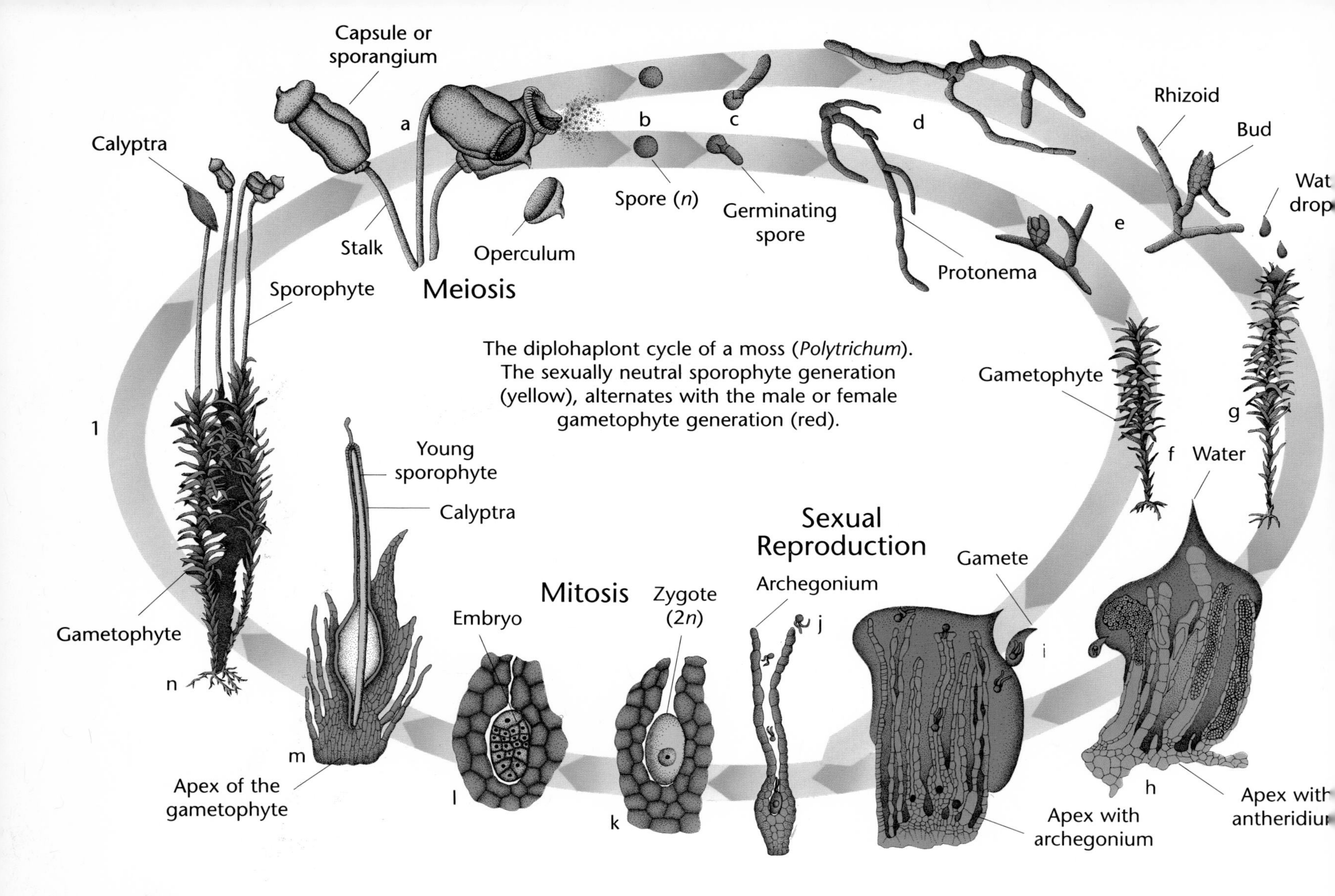

The development of the **embryo** gives rise to a sporophyte. The sporophyte looks completely different from the gametophyte and can carry out photosynthesis during the first part of its growth phase. The sporophyte is made up of a foot and a **stalk** with no branches. The stalk has a **capsule** at the top. Meiospores are produced in this capsule. In mosses, the capsules have a double wall that helps them to survive periods of drought.

The classification of mosses and hornworts is based on differences in the form of their sporophytes. The liverworts, on the other hand, are classified into two groups that differ according to the development of their gametophytes. The **anthocerotae** group may have been the ancestor of the simpler ferns. These hornworts have a small gametophyte, but their sporophytes have an indefinite growth period and have stomata. Stomata are openings in leaves and leaflike structures through which a plant exchanges gases with its environment. The sporophytes of anthocerotae can grow from the foot and make direct contact with the ground. In this way, the sporophyte can live independently of the gametophyte.

Miniature Ecosystems

Bryophytes live on all kinds of rocks and soils, but a forest is their ideal environment. In the forest, bryophytes develop in complex **associations** with many different **microorganisms**. The bryophytes and the microorganisms with which they live form true miniature **ecosystems**.

The largest moss species live near the equator. Their stems may be branched and many inches long. The inner structure of the stem is complex. Dead cells at the center of the stem are used to transport water. Other cells with rigid walls are interconnected by small holes. These cells look a lot like the **vascular** tissues of higher plants. Some moss species have many-layered leaflike structures that have a central **midrib**. The midrib is similar to a primitive vein. In contrast, the leaflike structures of liverworts exist in single layers.

1. In the maturing capsule (**a**) the spores (**b**) form through meiosis with haploid chromosomes (*n*). The spores germinate (**c**) and transform into protonema (**d**). Gametophytes (**f** and **g**) develop from the buds (**e**). The male gametophytes produce antheridia (**h**), which, when they are mature, produce male gametes (**i**). If water is present, the male gamete penetrates the archegonium (**j**) of the female gametophyte and fuses with the egg cell. This is a form of sexual reproduction. It produces the zygote with diploid chromosomes (**k**). The zygote divides by mitosis, forming the embryo (**l**) and then the young sporophyte enclosed in the calyptra (**m**). The adult plant (**n**) is made up of the gametophyte, which supports and feeds the sporophyte. **2.** and **3.** General and close-up views of *Conocephalum conicum*, a liverwort living in damp woodlands. The bubblelike stomata can be seen in the close-up view. **4.** The moss *Andreaea angustata*. **5.** The liverwort *Anthoceros*. **6.** The club moss *Lycopodium clavatum*. As plants evolved, the gametophyte (green) grew smaller, while the sporophyte (yellow) expanded.

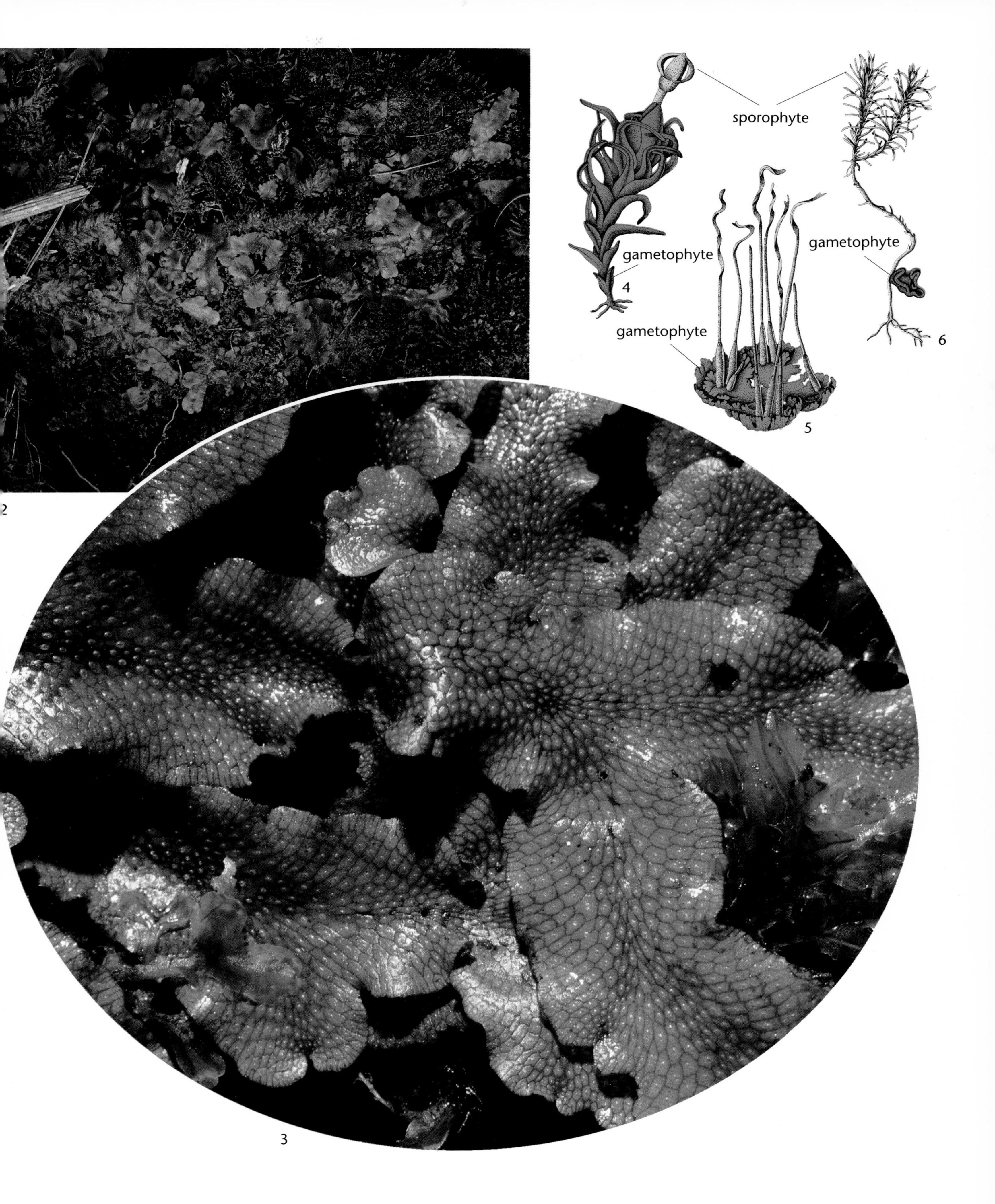
sporophyte
gametophyte
gametophyte
gametophyte
4
5
6
2
3

Ferns, Horsetails, and Club Mosses

Roots, Stems, and Leaves

Ferns, horsetails, and club mosses are classified in the pteridophyte group. Several new adaptations appeared with these plants. These adaptations allowed true land plants to develop. The most important adaptation was a reversal in the alternation of the generations. The sporophyte became the dominant generation. The gametophyte continued to become less important and has been reduced to only a few cells in modern flowering plants. However, among ferns, the gametophyte still has an independent role.

Water is needed for fertilization to take place in ferns. The union of the gametes produces a zygote. As it develops, the zygote changes to an embryo, then a **seedling**, and finally to an adult plant.

As early as the embryo phase, the sporophyte of ferns shows the development of the three systems of higher plants: roots, stems, and leaves. This **differentiation** occurred through evolution as tissues took on functions of support, transport, protection, storage, and photosynthesis. The **meristem tissues** are of major importance. They are the source of **primary growth** at the tip of the roots and the stem. The meristem also allows the **secondary growth** that thickens the plant.

1

2

1. Ferns of the species *Pteridium aquilinum* in the forest undergrowth. Ferns may also live in fairly dry environments as long as they have some shelter from direct sunlight. **2.** *Asplenium adiantum nigrum* showing the sori on the underside of the fronds. **3.** In the species *Dryopteris marginalis,* the sori are protected by a membrane called the indusium. Early fernlike plants were the ancestors of seed plants. The evolutionary success of the ferns is shown by the existence of more than 10,000 species, living in diverse habitats on every continent.

From *Rhynia* to the Ferns

Rhynia is one of the oldest land plants. This plant dates back to the early Devonian period, 345 to 395 million years ago. Today *Rhynia* is extinct.

Among today's plants, there are several species that originated at the end of the Devonian period. Among them are the horsetails and the club mosses. These plants are **herbaceous**—they have a nonwoody stem that dies each year. However, genera such as *Lepidodendron* and *Sigillaria*, which grew during the Carboniferous period, 280 to 345 million years ago, were trees. These plants formed huge forests and grew as high as 40 meters (131 feet).

Modern ferns are the group within the Pteridophyta division that evolved most recently. Ferns are usually found in humid environments and in forest undergrowth. They often have feathery leaves called fronds. Some ferns have developed unusual traits, such as climbing stems. Others have adapted to survival in dry conditions. In some tropical environments, tree ferns still survive.

The Diploid State and Heterospory

The evolution of the sporophyte was probably possible because of its diploid state. In this state, the extra genetic material makes it easier for the organism to adapt to changing environmental conditions. These adaptations may occur during the development of the plant. If you look at the range of pteridophytes, from the most primitive to the most recent, you see an increasing tendency to protect the spores. The spores are the cells that can produce new individuals withou fusing with other cells. They are formed i **sporangia.** In the earliest pteridophytes, spo rangia were located at the tip of the plant. I the horsetails, which evolved later, the spo rangia are located under the leaves, near th edges. In modern ferns, the sporangia ar grouped in structures called **sori**. Sori ar sometimes protected by a membrane calle an **indusium.** In some rare cases, they ar protected by a fold in the leaf that forms primitive **fruit** called a **sporocarp.**

Differentiation into smaller and large spores in the same plant was a major stage i the evolution of the pteridophytes. Thes spores are known as **macrospores** an **microspores.** Macrospores develop in game tophytes that carry archegonia and ar considered female. The male microspore

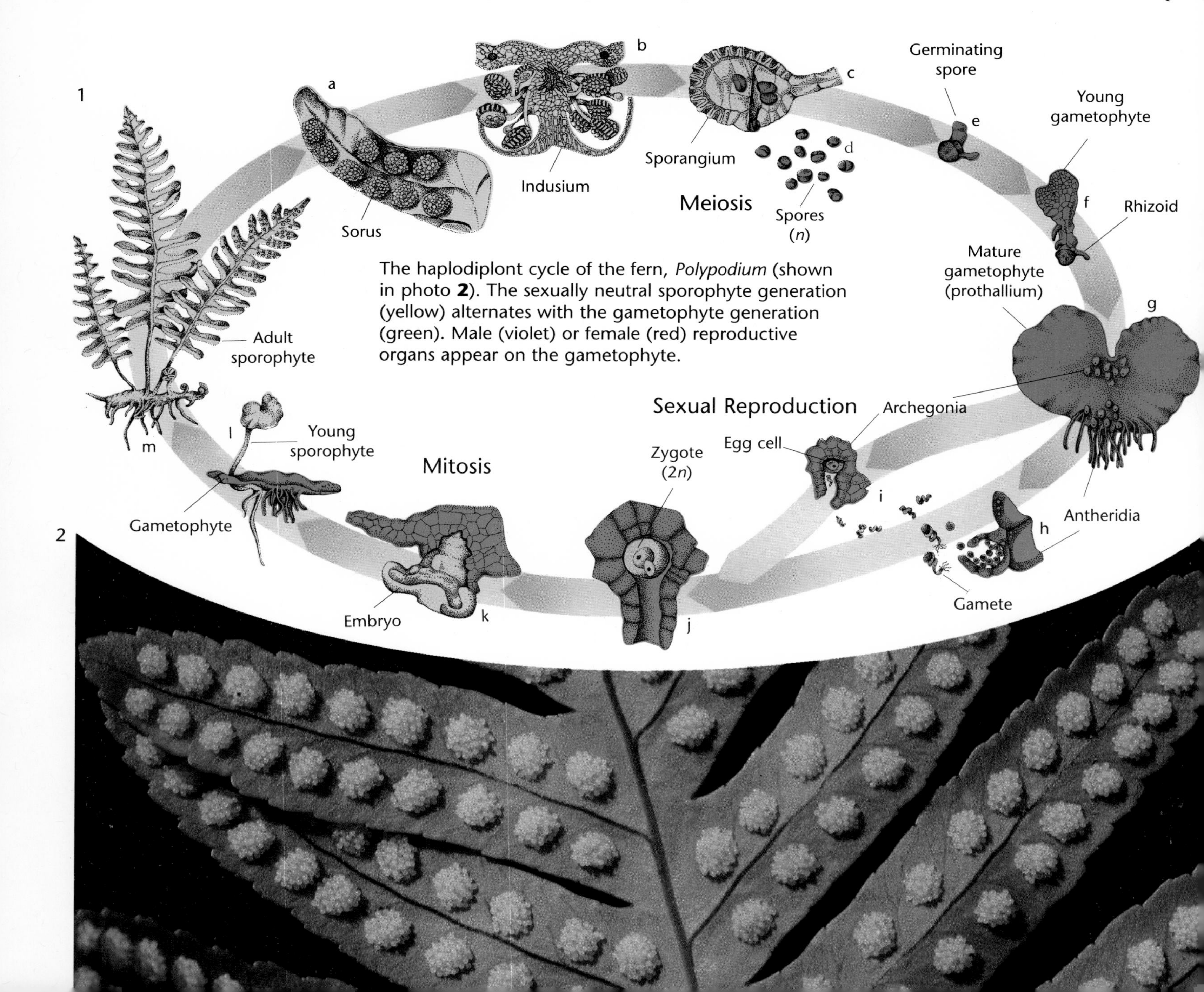

The haplodiplont cycle of the fern, *Polypodium* (shown in photo **2**). The sexually neutral sporophyte generation (yellow) alternates with the gametophyte generation (green). Male (violet) or female (red) reproductive organs appear on the gametophyte.

develop gametophytes carrying antheridia, which produce male gametes. The evolution toward **heterospory**, the production of more than one type of spore, probably took place in pteridophyte species that are now extinct.

Heterospory may have led to the evolution of the first flowering plants. Other changes occurred in plants as well. These changes include the evolutionary movement toward the sporophyte's becoming the main stage of the cycle and toward more protection for the macrospore. The spore's germination and the delay of its detachment from the main plant were also important adaptations. The ferns that have survived to the present day usually produce **isospores**. Isospores are spores that look the same and have no sexual differences.

1. Description of life cycle (shown page 26). **a.** The bottom of a fertile leaf densely covered with sori. **b.** A section of a *Dryopteris filix-mas* leaf with a sorus protected by an indusium. **c.** A sporangium in which spores (**d**) with a haploid chromosome set (*n*) form through meiosis. **e.** Following dispersal, the spores germinate and produce a gametophyte. **f.** The fern passes through a young stage, then a heart-shaped adult stage (**g**). The gametophyte carries the male antheridia (**h**) and the female archegonia (**i**). **j.** A zygote with diploid chromosomes (2*n*) formed through mitosis, develops into an embryo (**k**) and then a sporophyte (**l**), which is fed by the gametophyte. **m.** When the sporophyte becomes self-sufficient by rooting in the ground, the gametophyte dies.

3. *Equisetum maximum* on a forest floor.
Pages 28–29:
A close-up reconstruction of an imaginary forest from the Devonian period. The plants are between 10 centimeters (4 inches) and 1 meter (3.3 feet) high. Arthropods, perhaps the oldest land animals, including a scorpion, a springtail, and a myriapod with a segmented body, are shown. Fossils from this period offer a view of diversified primitive vascular vegetation. Some plants, such as *Drepanophycus spinaeiformis* and *Asteroxylon mackiei*, already look similar to the club mosses. Others, such as *Cooksonia caledonica*, *Rhynia major*, and *Zosterophyllum*, are very primitive and are considered to be the oldest land plants. They became land plants around 400 million years ago, between the late Silurian and early Devonian periods.

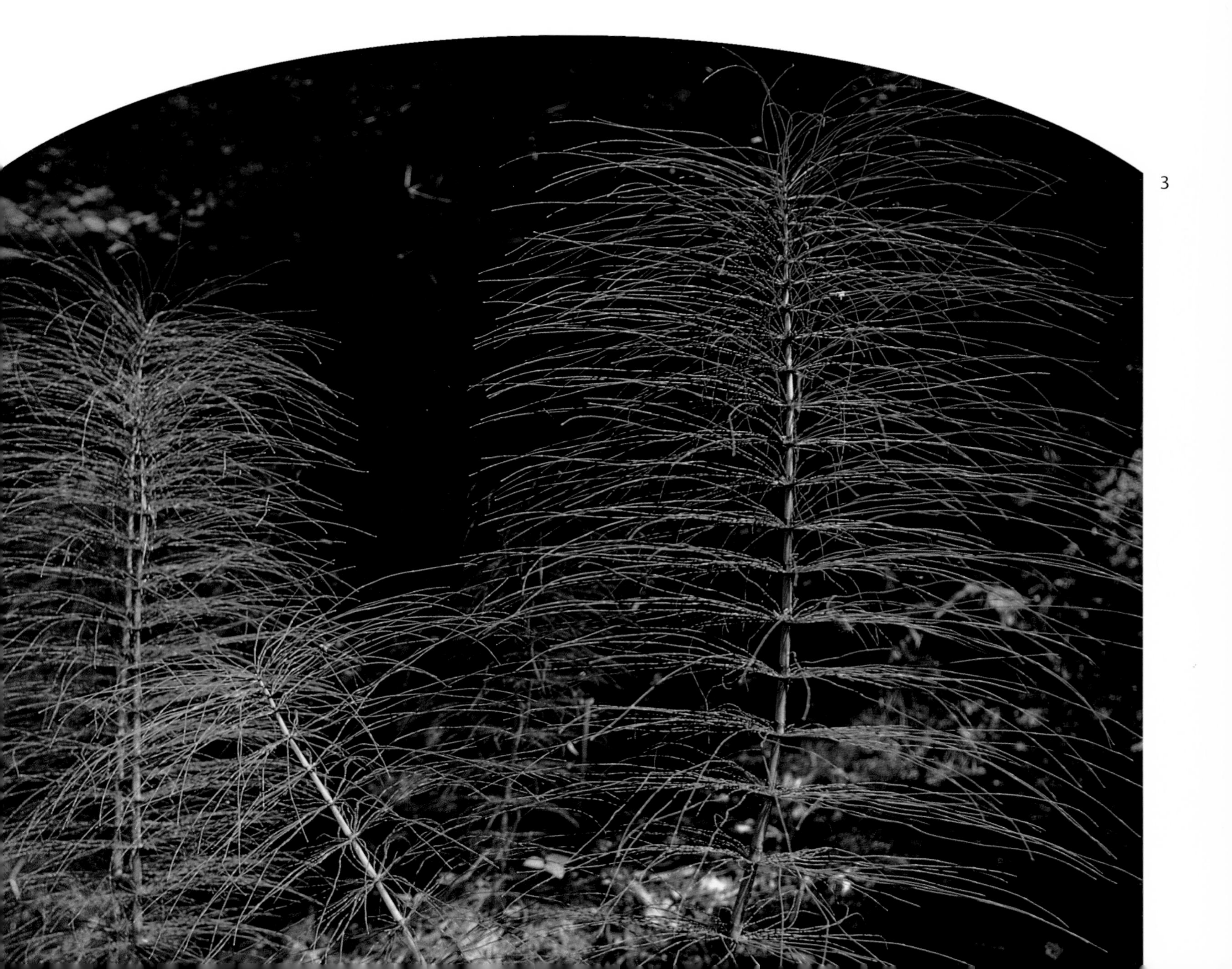

3

Drepanophycus spinaeiformis
Asteroxylon mackiei
Rhynia major
Springtail
Cooksonia caledonica
Scorpion

orneophyton lignieri
Psylophyton cernulatum
Pertica quadrifaria
Myriapod
Sciadophyton
Zosterophyllum
Rebuchia ovata

Seed Plants

1

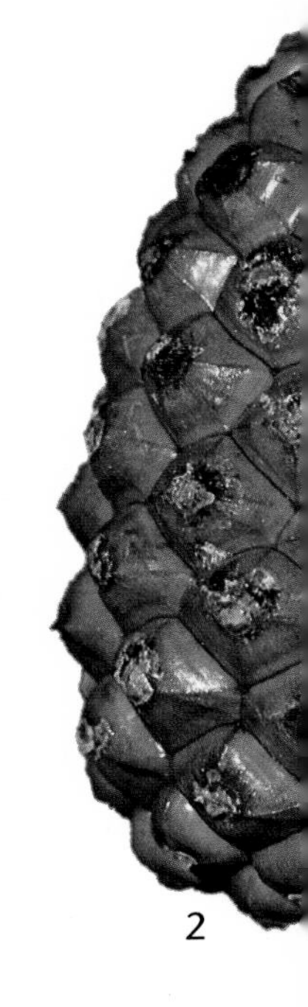

2

Seed Producers

As plants evolved, their reproductive structures became better protected. This tendency resulted in the formation of **seeds** as a protected means of reproduction that no longer needed water. The first plants to develop seeds were the seed ferns. These plants, which are now extinct, eventually gave rise to the higher land plants. The ancient seed ferns, such as *Medullosa*, formed from spore-producing ferns and flourished along with horsetails and club mosses. These plants existed in swampy forests during the Carboniferous period, 280 to 345 million years ago, and the Permian period, 225 to 280 million years ago, the last period of the Paleozoic era.

During the Mesozoic era, major climate changes led to higher temperatures and drier environments. Most tree ferns became extinct. Seed plants, ranging from very primitive examples to those with **flowers**, evolved. These plants, of the Spermatophyta division, first appeared during the Jurassic period, 140 to 195 million years ago. They spread rapidly during the Cretaceous period, 65 to 140 million years ago.

The modern seed plants include the **gymnosperms** and the angiosperms. The best-known gymnosperms are the conifers, which have naked seeds. The angiosperms are flowering plants that have seeds enclosed inside a protective structure, called the fruit.

The advantage of a seed is that it contains an embryo. This embryo is fed in the earliest stages of development by a food store inside the seed. The embryo is protected by one or more hard coats. This is a very efficient and independent adaptation for plants living on land environments.

The production of seeds is a complex process carried out by specialized organs. The presence of an **ovary** is the main feature that distinguishes the angiosperms from the gymnosperms. The gymnosperms are the more primitive seed plants. The **ovules**, the female **reproductive organs** from which the seeds develop, are exposed and not protected by an ovary. The more highly evolved angiosperms have ovaries within their flowers. These ovaries contain the ovules. When the ovaries are mature, they become the fruits that contain the seeds.

Seeds have a better food store in angiosperms than in gymnosperms. Angiosperm seeds are also more easily **dispersed**, or moved around, so that they can produce new plants. This explains why angiosperms are much more numerous and found in many more areas than all the other plant groups.

The Spermatophyte Life Cycle

The life cycle of the spermatophytes can be divided into two phases. First there is the **vegetative phase**. In this phase, the plant develops all the organs needed for sexual maturity, including the roots, the stem, and the leaves. In **annual** plants—those that germinate, grow, and die within one year—the vegetative phase lasts one growing season.

The second phase of the life cycle is the **reproductive phase.** In the reproductive phase, the plant uses its energy to develop flowers and reproduce sexually. A mature **perennial** plant, which lives for more than two growing seasons, can flower once a year. This increases its chances of reproducing. Genetic traits, usually coming from different plants, are combined during sexual reproduction. The resulting embryo is enclosed within a seed.

To ensure that offspring are produced, it is not sufficient to produce many seeds that are ready to germinate. It is also necessary for the embryos to have the right conditions in which to flourish after they are dispersed. If all seeds simply fell to the ground at the foot of the parent plant, they would have little chance of survival. They would soon be suffocated by the competition for food, water, and light. Plants have evolved systems to prevent this from happening. These systems permit the process of **seed dispersal**, the scattering of seeds to new areas.

Many species scatter their seeds a certain distance from the parent plant. This is a mechanical system of seed dispersal. An indirect system of seed dispersal is the production of soft, juicy fruit that is eaten by animals. Such seeds are dispersed involuntarily by the animals that eat the fruit. Other types of fruits and seeds have hooks, spines, hairs, or other adaptations to help disperse them in suitable environments.

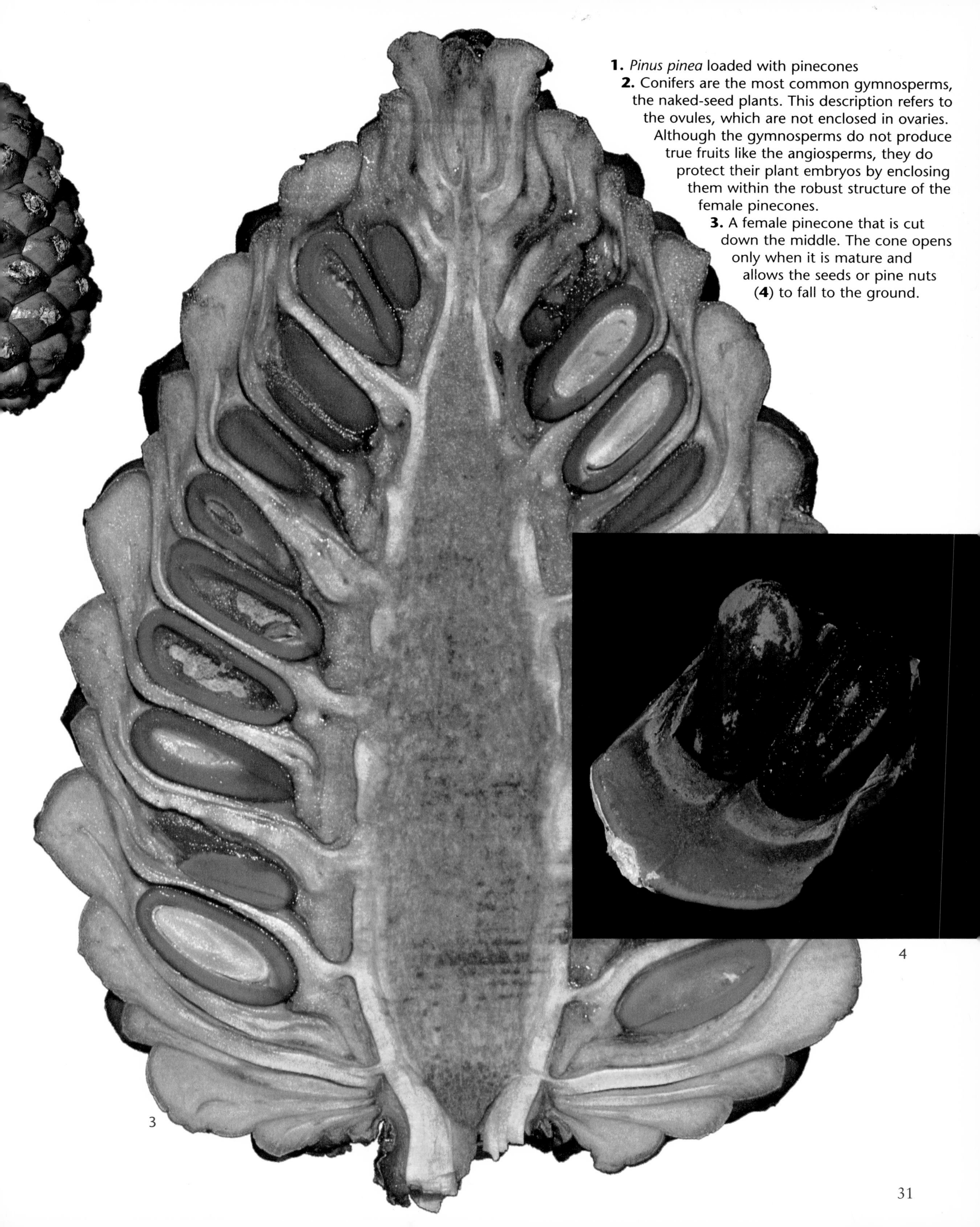

1. *Pinus pinea* loaded with pinecones
2. Conifers are the most common gymnosperms, the naked-seed plants. This description refers to the ovules, which are not enclosed in ovaries. Although the gymnosperms do not produce true fruits like the angiosperms, they do protect their plant embryos by enclosing them within the robust structure of the female pinecones.
3. A female pinecone that is cut down the middle. The cone opens only when it is mature and allows the seeds or pine nuts (**4**) to fall to the ground.

3

4

Wild fruits and cones use different methods of seed dispersal.
1. A sectioned fruit of *Liriodendron tulipifera* with winged seeds for wind dispersal. **2.** The wind-dispersed feathery fruit of *Clematis vitalba*, a climbing plant. **3.** The cone-shaped infructescence of *Magnolia grandiflora*, which produces bright red fruits. The magnolia may be the oldest angiosperm. **4.** The hard capsule with the nutlike fruit of the horse chestnut, *Aesculus hippocastanum*.
5. An immature cone of *Cupressus arizonica* cut across the middle. The cones of the cypresses are hard and round. When mature, they become woody, and the scales open, allowing the seeds with their narrow, winged membranes to fall out.

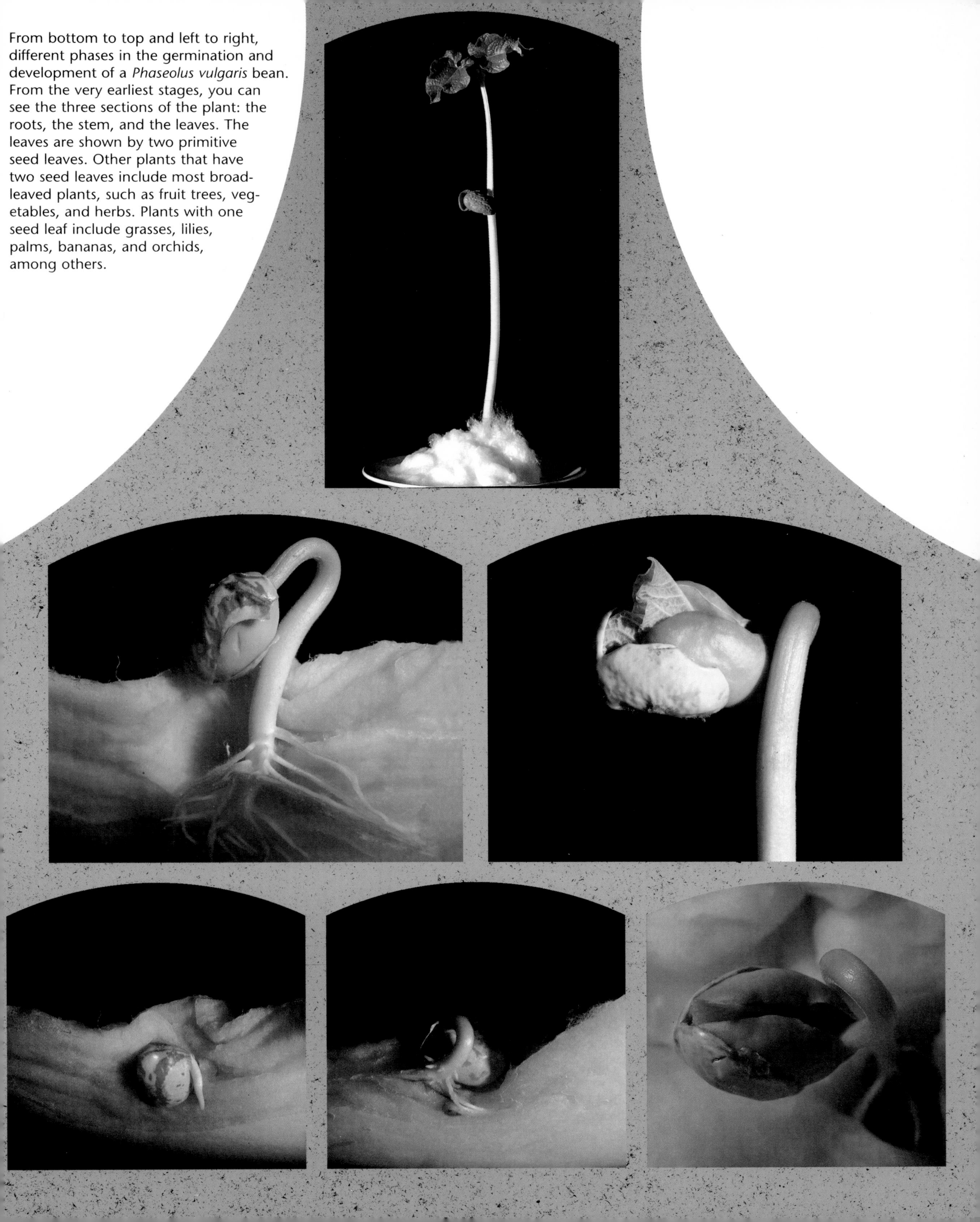

From bottom to top and left to right, different phases in the germination and development of a *Phaseolus vulgaris* bean. From the very earliest stages, you can see the three sections of the plant: the roots, the stem, and the leaves. The leaves are shown by two primitive seed leaves. Other plants that have two seed leaves include most broad-leaved plants, such as fruit trees, vegetables, and herbs. Plants with one seed leaf include grasses, lilies, palms, bananas, and orchids, among others.

Compared with the Devonian forest shown on pages 28–29, this forest from the Carboniferous period, 280 to 345 million years ago, is majestic in appearance. It is essentially composed of immense ferns, horsetails, and club mosses. Many of the plants tower over 20 meters (66 feet) above the forest floor and have woody stems. When these stems fossilized, they led to the creation of huge deposits of coal. In the foreground on the left are the trunks of the club mosses *Lepidodendron* (**1** and **9**) and *Sigillaria* (**3**, **4**, and **10**). The leaves of a herbaceous fern (**11**) can be seen together with the stems of spore-producing tree ferns like *Psaronius* (**5**) and seed-producing tree ferns such as *Medullosa* (**6**). On the right are the characteristic fronds of the giant horsetails *Calamites* (**7**). The trunk and foliage can be seen of the cone-bearing *Cordaitales* (**2** and **8**), the extinct ancestors of the conifers. A huge *Meganeura* dragonfly (**12**) flies over the swamp.

The scene is very different during the Cretaceous period, 65 to 140 million years ago. The plants have further developed to resist dry conditions, and the forest is generally more modern in appearance. Some of the most primitive plant groups have disappeared. Others, such as the horsetails and club mosses, have become less prominent. These plants no longer make a major contribution to the formation of forests. The ferns are more widespread, and new plant groups have appeared. In this reconstruction you can see several primitive conifers (**1**, **2**, and **3**), herbaceous ferns (**6**), tree ferns, such as the gigantic *Tempskya* (**4**), the small *Nathorstiana* (**7**), and *Cycadeoidea* (**5**). This was the era of the dinosaurs, but they were destined to become extinct by the end of the Cretaceous period. A bactrosaurus (**8**) can be seen running rapidly toward the shelter of the forest.

THE STRUCTURE OF THE FLOWER

The reproductive organs of the higher plants are called flowers. Flowers are modified and highly specialized leaves.

The Flowers of the Gymnosperms

The flowers of the gymnosperms are small and lack the bright colors or scents that attract insects. These plants have no need to make their flowers attractive because they are pollinated by the wind. A large amount of pollen, the male gamete, is produced to ensure that at least some grains reach the female flowers.

In the Middle Ages, conifer forests covered much of Europe. During the pollen season, everything was covered with a dusting of what was mistaken for a "rain of sulfur." In fact, the "sulfur" was masses of yellow pollen grains released by the conifers.

To avoid **self-pollination**, the fertilization of an ovule by pollen from the same plant, gymnosperms produce male and female flowers on the same plant that mature at different times. Self-pollination is undesirable because it produces weak offspring that are less adaptable.

The flowers of conifers, the best-known species in the gymnosperm group, are either male or female. In the spring, firs, pines, larches, and other conifers develop flower groups called **inflorescences**. These structures are made up of a long central stem on which a spiral of fertile leaves develops. The leaves are called **microsporophylls** and **macrosporophylls**. They carry the **pollen sacs** in the male flowers, and the ovules in the female flowers.

The Flowers of the Angiosperms

Angiosperms have flowers that are made up of more complex and specialized structures than the flowers of gymnosperms. The angiosperms have different systems of **pollination**.

Flowers have some structures that have many functions and a fertile structure for reproduction. The parts of a flower are arranged on a stem called a **peduncle**. The end of this stem is enlarged to form the **receptacle**. The outer parts of the flower are called **sepals** and are usually green. A ring of sepals forms the **calyx**, which protects the flower bud and acts as a support when the flower opens. The **corolla** is formed from one or more rings of **petals** set inside the calyx.

Flower petals are often brightly colored. Together, the calyx and corolla are called the **perianth**. The perianth attracts pollinating animals and protects the fertile structure. When the perianth is not made up of distinct petals and sepals, the flower is formed from **tepals**, as in the case of lilies.

The male fertile leaves, or microsporophylls, are modified to form the **stamens,** which together form the **androecium**. Each stamen is made up of a long, slim part called the **filament**. The filament has an enlarged tip that encloses the **anther**. Pollen grains are made by the anthers.

The female fertile leaves, or macrosporophylls, are also highly modified. They take the form of **pistils**, which together form the **gynoecium.** The pistil is a hollow organ made up of a swollen base, the ovary, and a long part called the **style**. The tip of the style expands to form the **stigma**.

The ovules that gather the fertilizing grains of pollen develop in the ovary. The union of the genetic material in the ovary leads to the development of seeds. A seed is a plant embryo with its food store. Depending on the species of plant, each ovary may contain from one to hundreds of ovules. Plant ovaries, therefore, produce different numbers of seeds.

Many flowers are not isolated structures. They may be grouped to form inflorescences. The inflorescences may be in the forms of spikes, as in wheat; clusters, as in vines; umbrellas or **umbels**, as in elders; or **capitula**, as in daisies.

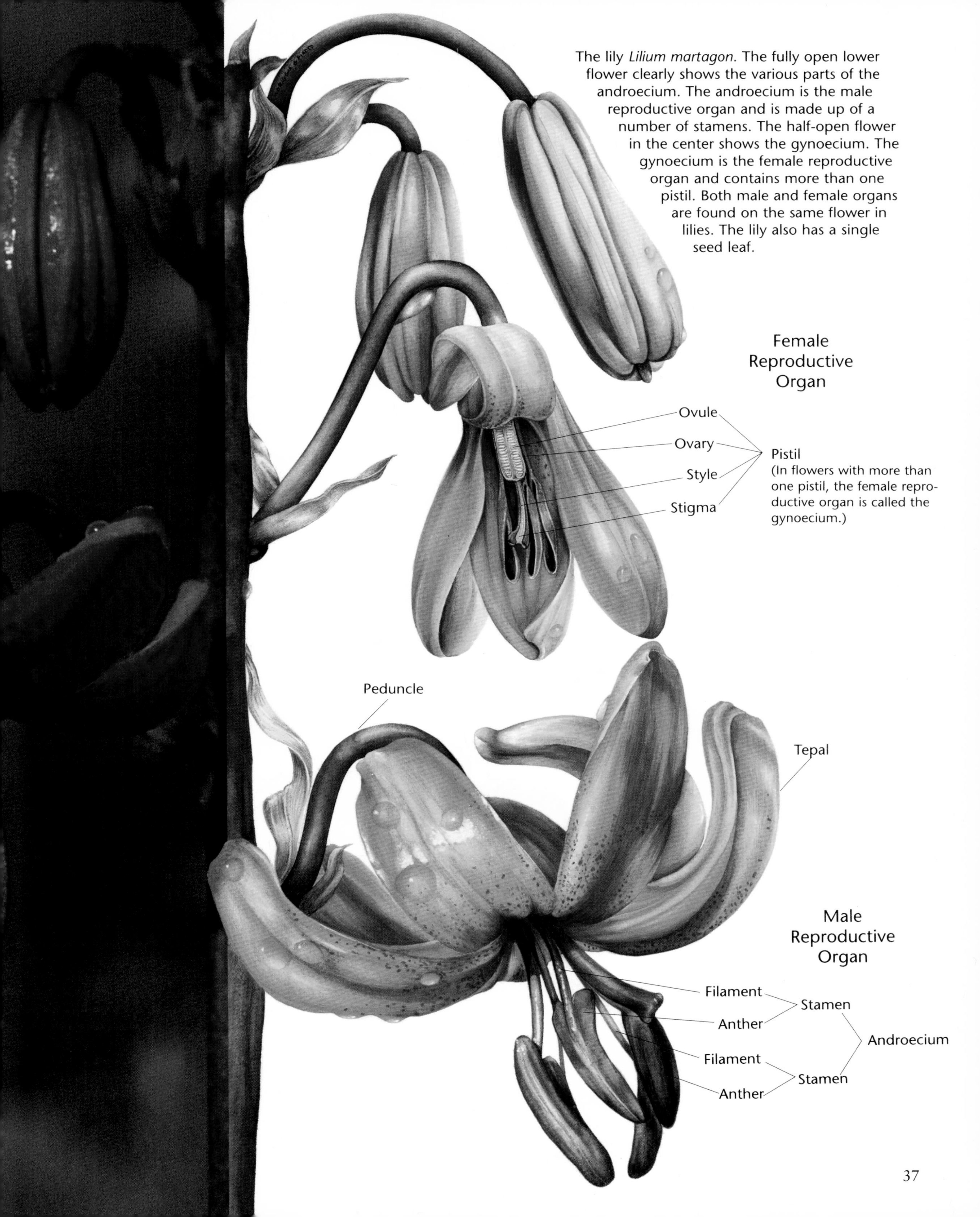

The lily *Lilium martagon*. The fully open lower flower clearly shows the various parts of the androecium. The androecium is the male reproductive organ and is made up of a number of stamens. The half-open flower in the center shows the gynoecium. The gynoecium is the female reproductive organ and contains more than one pistil. Both male and female organs are found on the same flower in lilies. The lily also has a single seed leaf.

The flower is the most highly evolved structure developed by plants to free them from the need for water as the agent of fertilization. In the gymnosperms, which are all wind pollinated, flowers are relatively simple. They are of one sex, with separate male and female flowers developing and maturing at different times on the same plant. The male flowers, usually located at the tip of a branch, are made up of pollen-producing scales.
1. A pollen-rich mature male *Pinus pinea* flower. **2.** Male *Cupressus arizonica* flowers.

3

1

4

5

6

7

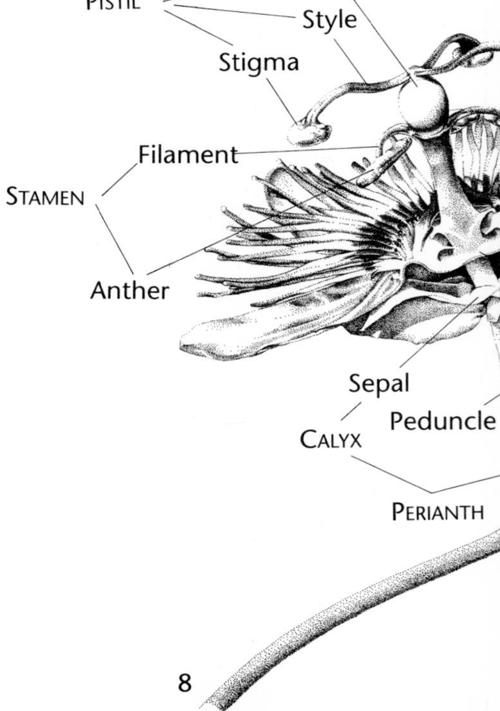

8

2

9

10

Flowers of angiosperms are much more complex and varied than those of gymnosperms. **3.** In the periwinkle, *Vinca*, the flower stem divides into five curved counterclockwise lobes. **4.** In the water lilies, *Nymphaea*, you see four green sepals below the many white, red, or yellow petals, depending on the species. **5** and **6.** In the sunflower, *Helianthus annuus*, and other composite flowers like the Transvaal daisy, *Gerbera jamesoni* (**10**), the capitulum, or central disc, is made up of many very small flowers set close together. The flowers are not all identical. The ones around the outer circumference have a large, tongue-shaped corolla, which is usually mistaken for a petal. **9.** In some flowers, like the marsh marigold (*Caltha palustris*), the true petals may be replaced by a calyx with the size, appearance, and color of a corolla. **11.** In the African violet, *Saintpaulia ionantha*, the corolla has two petals that are smaller than the others. **7.** In the orchids, one of the three petals has evolved to act as a landing pad for the pollinating insects. The petals have signaling colors that are lighter at the center, so they can be seen from a distance. **8.** The flower of the sectioned passionflower, *Passiflora coerulea*, has an anatomy that clearly shows the different parts of a plant with two seed leaves.

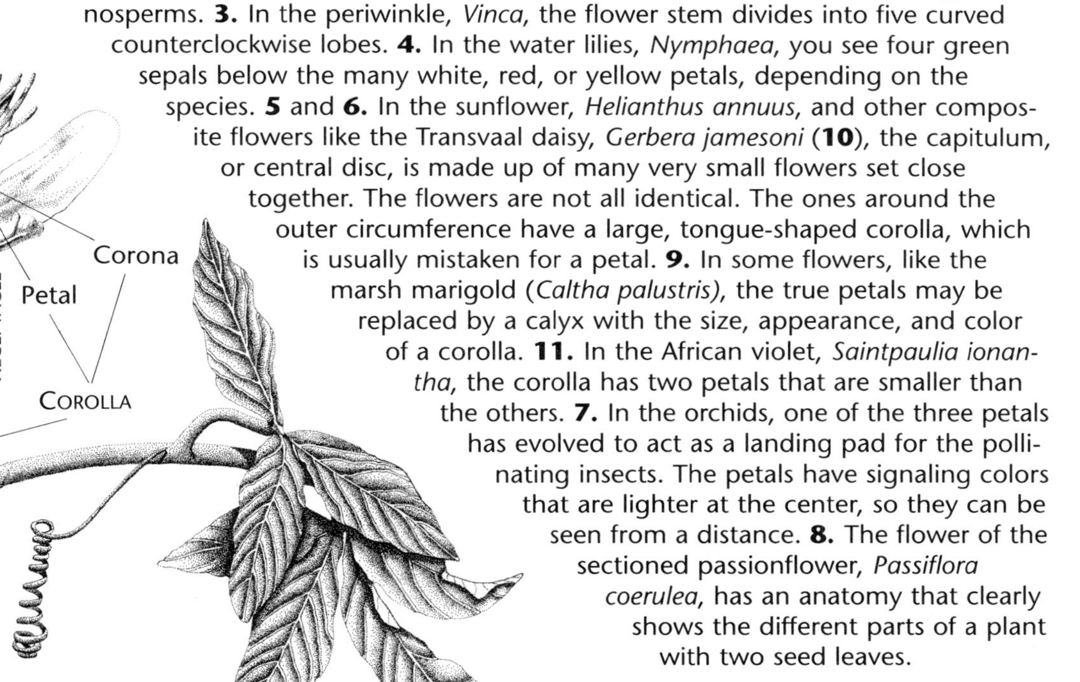

11

Pollination and Fertilization

Attracting Animals

In angiosperms, pollination is the transport of pollen from the anther of one flower to the pistil of another. Many angiosperms rely on the wind for this task. However, others have developed systems that make use of other agents. These agents often involve animals such as insects, birds, or bats.

The most dramatic examples of insect pollination involve flowers that attract males of only one species of insect by imitating the shape, color, and scent of the female insect. More often, the pollinating insect is attracted by the pollen or **nectar**. When the insect gathers the pollen or nectar for itself, some of the pollen grains collect on its body. When it visits the next flower, these grains are brushed off, fertilizing the second plant.

Certain species of birds transport pollen. Hummingbirds have developed specialized beaks and evolved parallel to the plants from which they get nectar. These birds are dependent on the plants for food. They are rewarded for their pollinating services with the sugary substances contained in nectar.

Self-pollination and Cross-pollination

In the reproductive phase of the flower, fertilization follows pollination. During fertilization, the female gamete, called the ovule, is united with the male gamete, called the microspore. There are two types of pollination: self-pollination and **cross-pollination**. In self-pollination, gametes from the same flower or the same plant unite. In cross-pollination, pollen from another plant fertilizes the ovule. Thus, **cross-fertilization** occurs only when the pollen has been transported some distance.

In general, self-pollination results in little genetic variation. So plants that self-pollinate are not very adaptable to new environmental conditions. They may also develop undesirable traits.

1

2

3

4

5

6

1. A fly on a daisy, *Bellis perennis*.
2. Scarab beetles on a *Leontodon*.
3. A *Lycaena* butterfly on a *Scabiosa*.
4. A moth on a sweet pea flower.
5. A caterpillar on a bramble *Rubus fructicosus*.
6. A burnet moth on a *Scabiosa*.
7. A caterpillar on a rose.
8. A *Tettigonia* larva on a *Hieracium*.

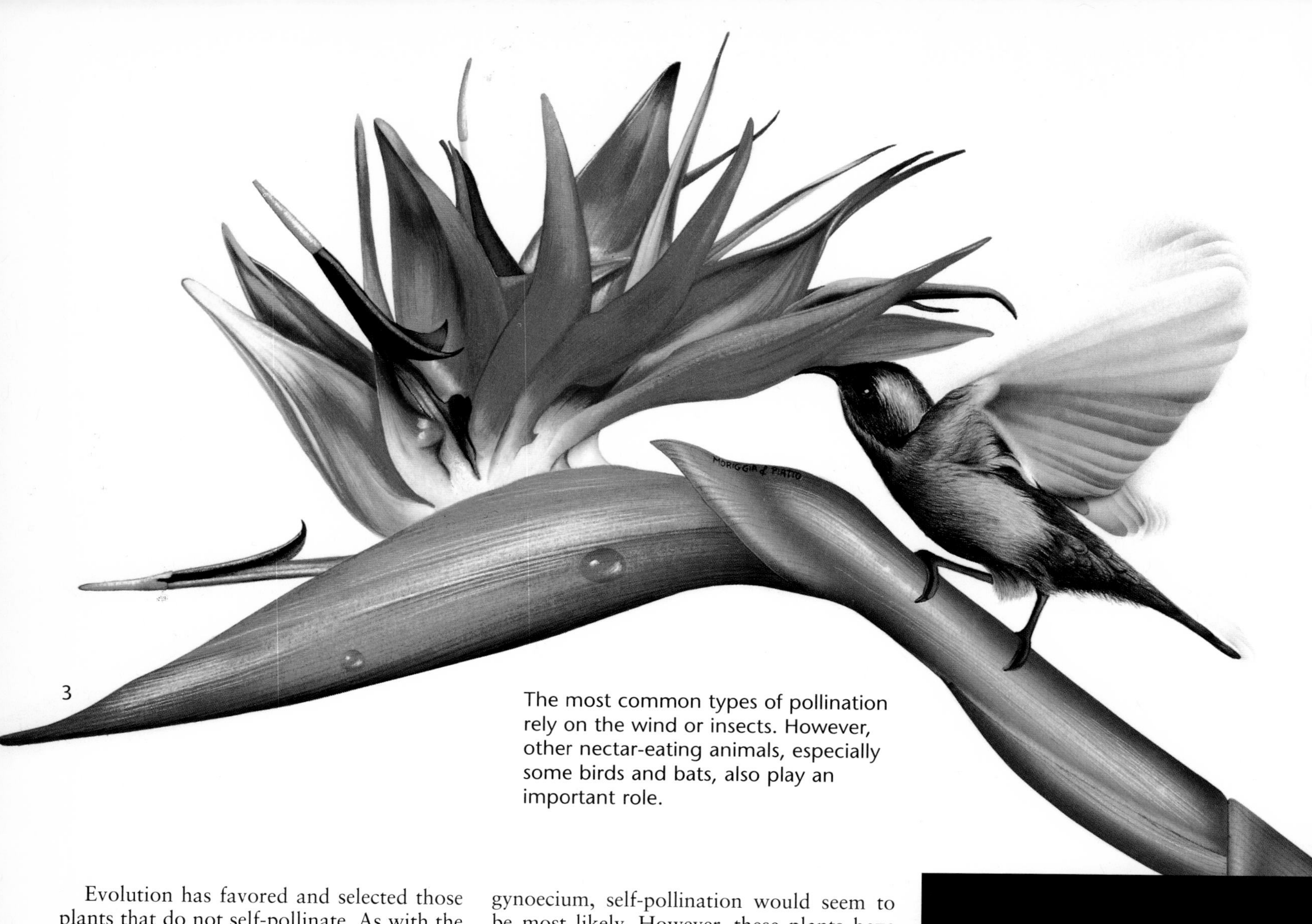

3

The most common types of pollination rely on the wind or insects. However, other nectar-eating animals, especially some birds and bats, also play an important role.

Evolution has favored and selected those plants that do not self-pollinate. As with the conifers, the most direct way to avoid self-pollination is to produce unisex flowers. These flowers are either **staminate** (male) or **pistillate** (female). Species that produce male and female flowers on different plants, such as willows and poplars, are said to be **dioecious.** Species that produce male and female flowers on the same plants, such as corn, are **monoecious.** Being monoecious is not always desirable. Monoecious plants use a lot of energy in their vegetative phase. Often this does not facilitate successful pollination.

Hermaphrodites

In angiosperms with **hermaphrodite** flowers, flowers having both an androecium and a gynoecium, self-pollination would seem to be most likely. However, these plants have evolved ways to avoid this. Often, the male and female parts of the flower may mature at different times. If at the end of the season none of the ovules have been fertilized, the flower may have to fall back on self-pollination, which is better than no pollination at all.

Many angiosperms cannot self-pollinate. Such species are said to be **self-sterile.** The pollen produced by all the flowers of an individual plant is incapable of fertilizing the ovules of the same plant. In this case, the species must cross-pollinate. It is probable that this system, together with other factors, made a major contribution to the evolution of the angiosperms.

1

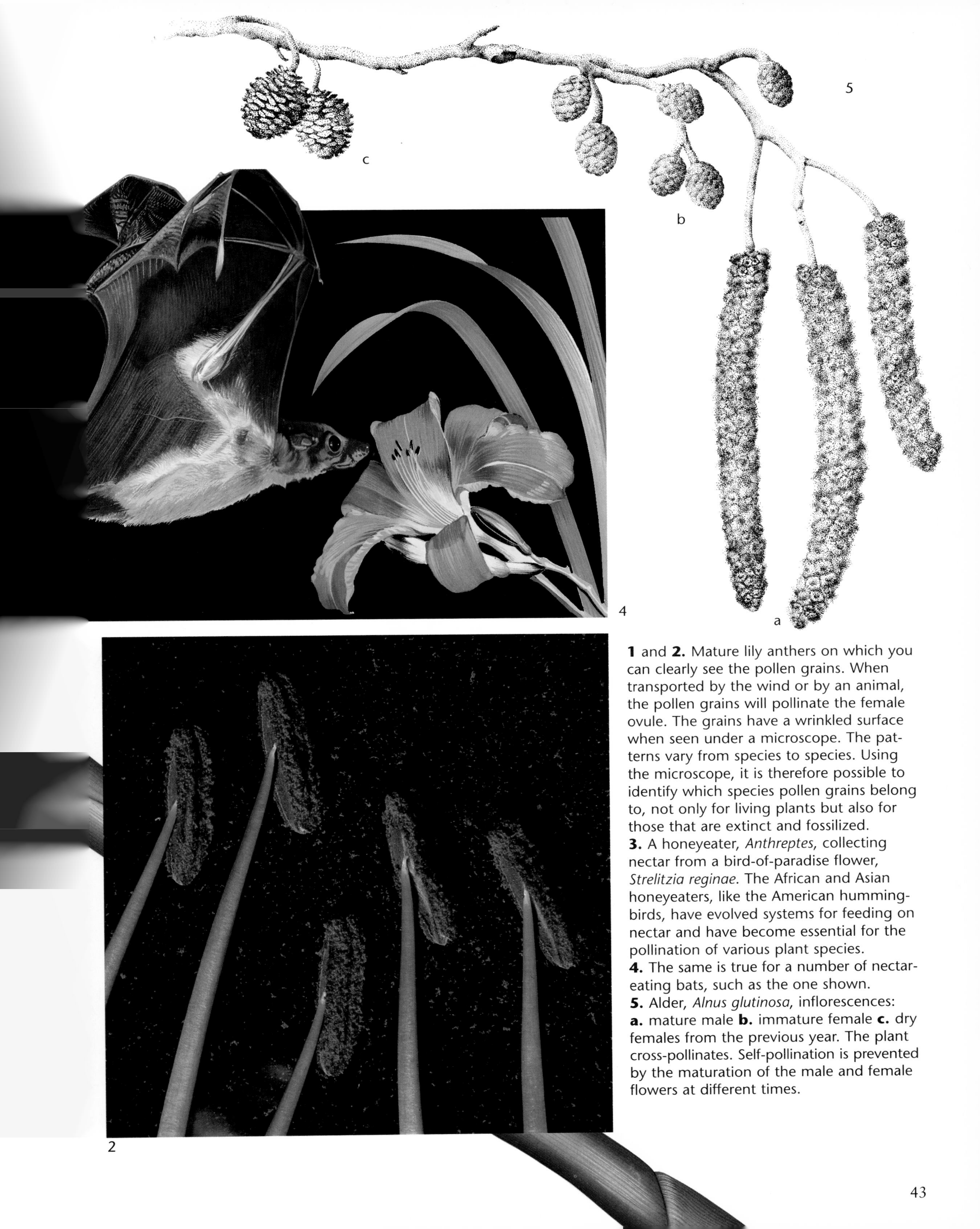

1 and **2.** Mature lily anthers on which you can clearly see the pollen grains. When transported by the wind or by an animal, the pollen grains will pollinate the female ovule. The grains have a wrinkled surface when seen under a microscope. The patterns vary from species to species. Using the microscope, it is therefore possible to identify which species pollen grains belong to, not only for living plants but also for those that are extinct and fossilized.
3. A honeyeater, *Anthreptes*, collecting nectar from a bird-of-paradise flower, *Strelitzia reginae*. The African and Asian honeyeaters, like the American hummingbirds, have evolved systems for feeding on nectar and have become essential for the pollination of various plant species.
4. The same is true for a number of nectar-eating bats, such as the one shown.
5. Alder, *Alnus glutinosa*, inflorescences:
a. mature male **b.** immature female **c.** dry females from the previous year. The plant cross-pollinates. Self-pollination is prevented by the maturation of the male and female flowers at different times.

Fruits and Seeds

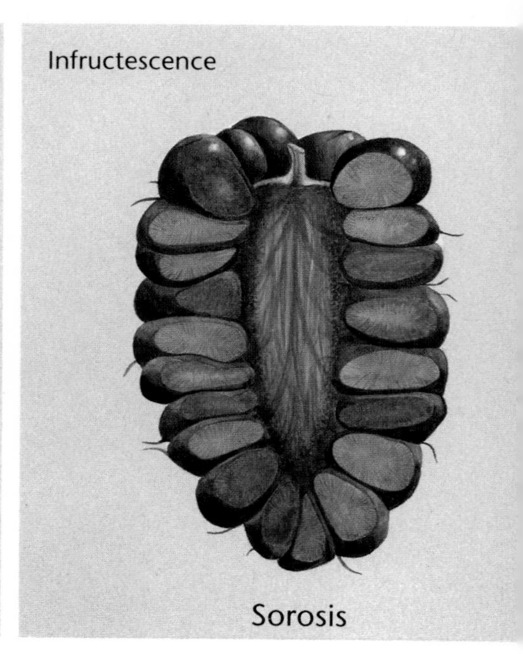

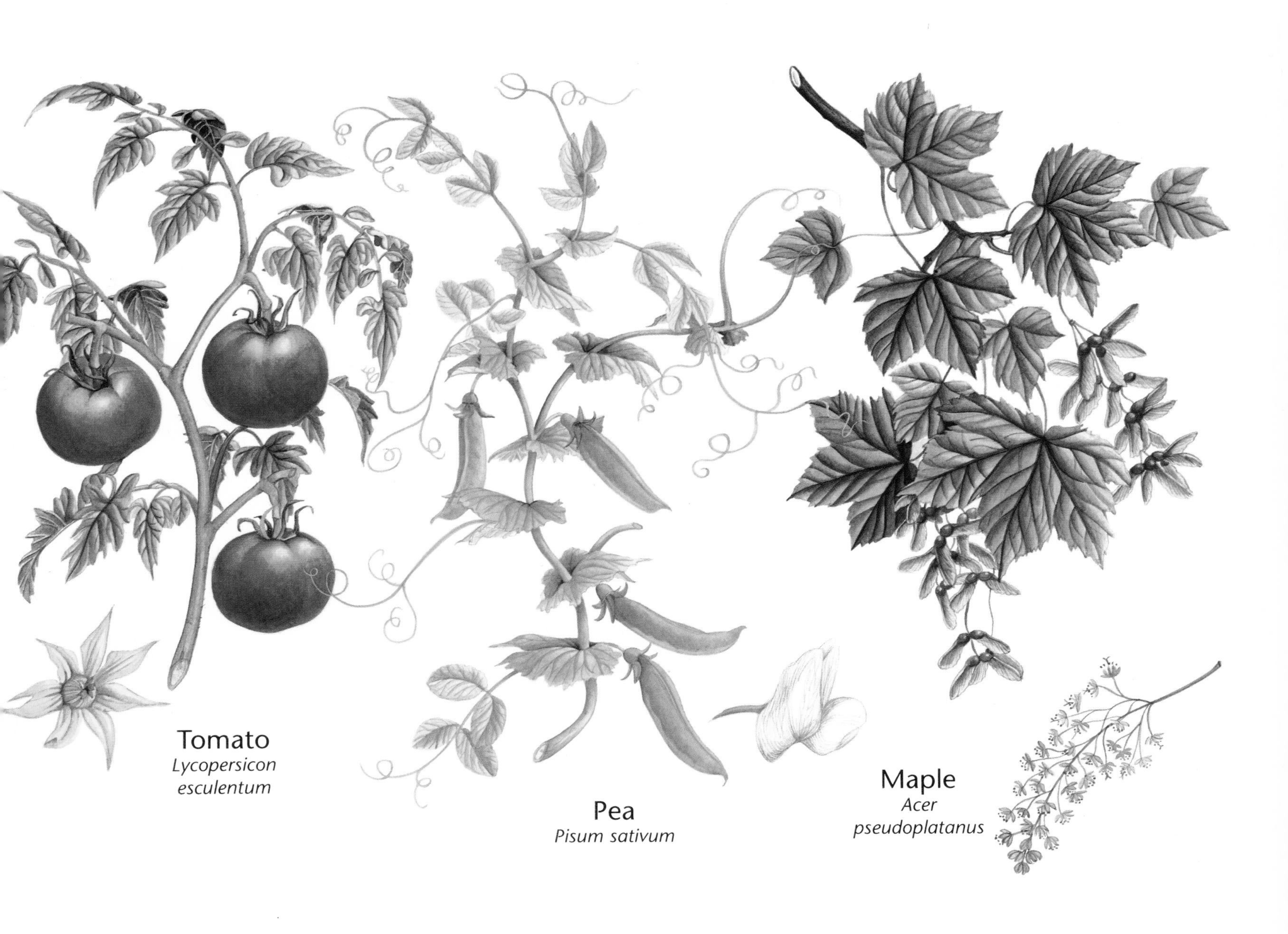
Tomato
Lycopersicon
esculentum
Pea
Pisum sativum
Maple
Acer
pseudoplatanus

leshy fruit
Berry

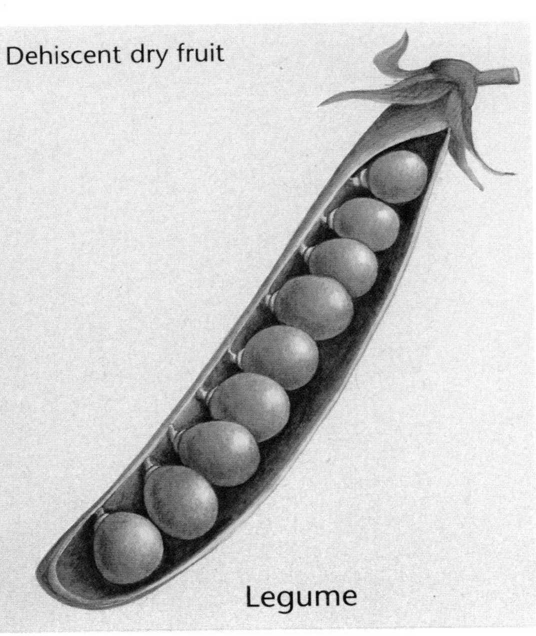
Dehiscent dry fruit
Legume

Indehiscent dry fruit
Winged Samara

Single and Multiple Flowerings

The role of flowers is completed after the ovule is fertilized. Then sepals, petals, and stamens wither, and the plant uses its energy to develop the ovary and the ovule it contains. Within a few weeks, the ovary and its seeds are mature. In many plants the fruit also incorporates other modified parts of the flower, such as the receptacle and the calyx. The fruit is a mature ovary with or without other parts of the flower. A seed is a mature ovule contained within the fruit.

Many plants die after they flower and produce fruit. Species that flower and fruit only once are said to be **monocarpic**. All annual plants are monocarpic. Annuals live for only one growing season. Some **biennials**, plants having a two-year growth cycle, are also monocarpic. An example of a biennial is the water hemlock, an herb. During the first year of their lives, biennials accumulate nutrient stores. In the second year, they flower and fruit. Finally, even though they may live for many years, some perennial plants wither and die after flowering. An example of such a plant is the agave.

In contrast, mature **polycarpic** plants flower and fruit periodically according to the seasons. Trees and shrubs are polycarpic. In this kind of plant, the vegetative stage may be brief, as in the case of many herbaceous plants, or it may last for a long time and coincide with the reproductive stage, as in woody plants.

Types of Fruits

The fruits of angiosperms have many different structures, shapes, and colors. Many trees, such as walnuts and plums, produce fruits containing a single seed. Such plants are called **monosperms**. Other types of fruits, such as capsules and **legumes**, contain many seeds. Plants that produce fruits that contain many seeds are called **multisperms**.

Fruits are classified into two major groups: **fleshy fruits** and **dry fruits**. Fleshy fruits, such as **berries**, have a soft, fleshy outer part. **Drupes**, such as peaches, apricots, and plums, are fleshy fruits with a pit containing a single seed.

Dry fruits have a hard outer part. They may be either **dehiscent** or **indehiscent**. Dehiscent fruits open to release their seeds like the pods of the legumes. Indehiscent fruits stay closed to protect the seeds, such as the **samaras** of the maple.

There are also **false fruits**. Most of a false fruit forms from modifications to parts of the flower other than the ovary. **Pomes** are false fruits with a leathery core, which contains seeds and grows from the ovary or ovaries of a flower. The best-known false fruits are apples and pears. The fleshy parts of these fruits develop from the receptacle. The core is what remains of the ovary.

A blackberry is a type of fruit called an **aggregate fruit**. Such fruit forms from several ovaries of the same flower. If the fruit develops from an inflorescence, it may be an **infructescence**. Examples of these kinds of fruits include the **syconium** of the fig plant, the **sorosis** of the mulberry, and the strawberry, a false fruit.

Seed Dispersal and Germination

Fruits protect and disperse their seeds in various ways. Like pollen, seeds may be transported in different ways. Wind, water, and animals all help to disperse seeds. Before reaching a desirable environment, seeds are in a period of **dormancy**, or very slow growth. Even when they reach a suitable environment, the seeds are not yet ready to germinate. They must go through a **postmaturation** process. The seeds of the mistletoe and the holly go through one of the most unusual processes. Before they can germinate, these seeds must pass through the intestines of the birds that feed on them. The combination of the birds' body temperature and **gastric juices** triggers the mechanisms that lead to the postmaturation of the seeds.

Other seeds do not germinate until they come into contact with water, and then only at a certain temperature. Some seeds or fruits contain substances that prevent germination.

The fruits of the angiosperms vary widely in shape, color, and method of seed dispersal. The fruits are as varied as the flowers from which they derive. Here are a few examples of wild fruits on their plants.
1. The berries of *Smilax aspera*, a thorny climbing plant of the Mediterranean maquis.
2. The dates of the dwarf palm, *Chamaerops humilis*, are long berries similar to drupes.
3. *Pittosporum* capsules.
4. The hooked, capsulelike infructescence of the sweet gum tree, *Liquidambar styraciflua*.

3

4

The seeds germinate only when these substances have broken down. Generally, in regions with clearly defined seasons, such mechanisms are useful because they ensure that the seeds do not germinate just before the cold season, when the young plants are likely to die.

1

2

4

3

Two examples of cultivated fruit and the Mediterranean agricultural environments in which they grow. This kind of landscape has been typical of parts of Europe for thousands of years.
1. Terraced vineyards in Liguria, Italy.
2. The fruit of the vine is a berry called a grape. Grapes grow in bunches.
3. Olive groves in Tuscany, Italy.
4. The olive fruit is a drupe with a fleshy outer layer and a very hard fruit wall.

Angiosperms: Multiform Plants

Adaptability

A main feature of the angiosperms, and one that has helped them thrive, is their variety of shapes, sizes, and reproductive systems. This flexibility has allowed angiosperms to adapt to very different physical and chemical conditions. The genetic variations that have evolved over the ages have produced numerous types of plants. These include trees, shrubs, herbaceous plants, erect plants, climbers, creepers, floating plants, and underwater plants. Thus, angiosperms are able to colonize desert areas, wet regions, cold arctic plains, hot tropical areas, lowlands, and mountains.

Each plant is an individual whose adult form results from its environment and its genetic program. The trees that grow in a dense forest have a slimmer shape than those of the same species growing in isolation. The trees growing on the edge of a forest are very leafy, while those living in windy areas, such as on a coast or mountain uplands, have flattened leaves and bent stems that resist the force of the wind.

Diversity and Competition

The first plants to colonize an area usually have plenty of space in which to grow. Their presence creates **microclimates** suitable to

In Egypt, about 3,500 years ago, a noble had a painting showing a garden put on the wall of his tomb at Thebes. It shows that gardens were part of human culture in this society. Later, this tradition was combined throughout Europe with the Greek concept of an open space in which plants could be cultivated for pleasure.

other plant species. These new species may find shelter from the sun beneath the leaves of the original plants. At first, the vegetation is open and consists mostly of herbaceous plants. Over time, more dense and varied vegetation develops. Plants of the same and different species begin to compete for the nutrients in the soil, water, and sunlight.

Forest trees try to expose as much of their **canopies** to the sun as possible. In contrast, species that have adapted to shaded conditions flourish on the forest floor. The various species develop different root systems to help reduce competition. Some have root systems that spread in a horizontal direction. Others have roots that extend downward. Evolution has also resulted in plants that can release substances through their roots or leaves that are poisonous to other competing species.

Epiphytes and Parasites

Various plant species may become interdependent, especially those living in complex ecosystems. This happens with the **corticolous plants**, plants that climb up woody plants, and the trees that support them. The typical appearance of the equatorial forest is due to its many **epiphytes**, plants that grow on other plants. These species of corticolous plants, such as bromeliads and orchids, have adapted to this environment. Epiphytes have **aerial roots** and can absorb water and mineral salts through their leaves. Some epiphytes can live on many species of trees, causing almost no damage to the trees. Often, epiphytes depend on the chemicals produced by the plant that supports them.

Parasitism is a relationship in which one organism, the parasite, lives on or in another, the host. Often the parasite harms, but does not kill, its host. **Hemiparasites**, such as mistletoe, can carry out photosynthesis but also feed on the sap of their host plant. The mistletoe feeds through **haustoria**, branching filaments that penetrate the host plant. The true parasitism of species such as the dodder, *Cuscuta epithymum*, is even more damaging for the host plant. These parasitic plants are incapable of photosynthesis. They obtain all their food by penetrating their hosts with haustoria.

The relationships between plants and between plants and other organisms are governed by ecological principles. Relationships develop that range from symbiosis to parasitism.

1. An epiphyte cactus, *Epiphyllum* genus, growing on a palm. The cactus uses the palm only as a support.

2. Ivy, using special aerial roots, climbs up a tree trunk.

The Deep Green Planet

A flying saucer observing the Earth from many points—over the Amazon, Siberia, or Borneo, for example—would see a rolling sea of treetops. Forests are the fundamental environment of our planet. They are the environment that dominates and persists over time, following the colonization of Earth by living things. In a climate that is not too cold or dry, the result of colonization will be a forest of some kind. Whether deciduous broadleaf or evergreen, mixed or coniferous, tropical or temperate, it will still be a forest.

The difference between town and countryside is decided by humans. If nature was left to its own devices, the difference would disappear in a sea of trees, like the famous temples of Cambodia. Apart from the oceans, the tundra, and the deserts, the whole planet is, will be, or would be covered with forest. In a forest environment, the terrestrial ecosystems draw breath. Their continuous labors cease, and they finally rest in a stable, durable form that is resistant to change. The concept of the town and the countryside has no future, except in the hands of people who decide it on the basis of their needs. Forests are the true future of the planet. When they are felled, burned, and uprooted, it is that very future that is being destroyed.

Ecologists say that forests are climax environments. Climax environments are stable, well defined, and balanced with a wide range of different species, microclimates, and subsystems. The destruction of forests means destroying something that was established and lasting, and replacing it with something new, unstable, and temporary. The destruction of even small areas of a large forest means the extermination of native species. It also means the wiping out of small worlds with unique characteristics. The destruction of the world's great forests at the current rate—150,000 square kilometers (58,000 square miles) each year—means the destruction of the planet itself. It means increasing the danger of total collapse day by day. Whatever the outcome, our planet without its forests will be a different world: one languishing in the memories of its past splendors. Let's hope that terrible day is still a long way off, and better still, that it never arrives.

Renato Massa

GLOSSARY

adaptation A process in which a species changes with environmental conditions

aerial root A root that rises above the ground

aggregate fruit Fruit formed from several ovaries of the same flower

algae Single-celled and many-celled plantlike organisms with rootlike structures that photosynthesize. They are classified as protists.

alternation of generations A process in which a sexually produced generation alternates with an asexually produced generation

alternation of nuclear phase The alternation of the number of chromosomes from *n* to *2n* and vice versa. In diploid organisms, such as animals, the *n* phase is represented by the gametes, while the *2n* phase represents all individuals from the moment of fertilization. With haploids and diplohaplonts, the *n* phase begins with the production of spores through meiosis and continues with the development of the gametophyte through mitosis.

androecium The male part of a flower made up of one or more stamens

angiosperms Flowering plants whose seeds are formed within a special structure called an ovary, which matures to form a fruit

annual A plant that lives for only one growing season

anther In a flower, the part of the stamen where the pollen grains are produced

antheridium The male sex organ in bryophytes and pteridophytes that produces gametes

anthocerotae Hornwort group that may have given rise to simple ferns

archegonium The female sex organ in bryophytes and pteridophytes that produces gametes

association A plant community that covers a wide area and is made up of a certain population of species with a characteristic appearance and habitat and with a stable duration

berry A simple fruit, fleshy when mature and containing one or more seeds. Tomatoes, grapes, and cucumbers are berries.

biennial A plant that lives for two years

binary fission A kind of cell reproduction in which two identical cells develop from one original cell

bryophytes Nonvascular land plants, including the mosses and liverworts that lack true roots, stems, and leaves

calyx The external part of the flower formed from the sepals, which protects the flower bud and acts as a support when the flower opens

canopy The layer of thick vegetation formed by the branches and leaves at the tops of tall trees in a forest

capitulum An inflorescence of a flower in which each petal is a single flower, although all the petals together form a flower head

capsule In the bryophytes, the organ in which the spores are produced. In the angiosperms, a dry fruit containing many seeds that opens along one or more sides when mature.

chlorophytes Green algae

chloroplasts Tiny structures containing chlorophyll that are part of plant cells and enable plants to photosynthesize

chromosome A small, threadlike cell part that carries genetic information and is located in the nucleus

conjugation The transfer of genetic material between two single-celled organisms, such as bacteria, protozoa, and some algae and fungi

corolla Part of a flower formed from one or more rings of petals arranged within the calyx

corticolous plants Plants that live on the surface of bark

cross-fertilization Union of gametes from two different plants of the same species

cross-pollination Transfer of pollen from the stamen of a flower on one plant to the pistil of a flower on another plant

cuticle Covering over the outer layer of leaf cells

dehiscent Having an opening that, whe mature, releases seeds from a dry fruit

diatom Single-celled brown algae with silic shells in two parts

differentiation A process in many-celle organisms in which one cell develops int many different types of cells in the embryo

dioecious Having either male reproductiv organs or female reproductive organs in a individual plant

diplohaplont cycle A life cycle characterize by an alternation of haploid (gametophyte and diploid (spermatophyte) generations

diploid Organism with two sets of chromo somes *(2n)* that are created through sexua reproduction

diploid cycle A life cycle consisting of a sin gle diploid generation (*2n*) in which meiosi produces gametes rather than spores. Th gametes unite to form a zygote.

disperse To scatter

dormancy Period of inactivity during a lif cycle

drupe A simple fruit made up of an oute fleshy layer and a woody center, called th pit, containing a single seed. Peaches, plums and apricots are drupes.

dry fruit A simple fruit with a hard outer tis sue that develops from the walls of the ovary This external structure dries out as the frui matures.

ecosystem All the living and nonliving part of an environment that interact to produce stable system

egg A female gamete

embryo In seed plants, the young sporophyt (*2n*) before it begins a period of rapi growth

epiphyte A plant that lives on other plants but does not feed on them

false fruit A fruit, such as an apple, in whic the juicy flesh forms from modifications c parts of the flower other than the ovary

lament In flowers, the slim, long part of the amen that carries the anther

eshy fruit A simple fruit with soft, external ssue formed from the walls of the ovary that mains soft and juicy as the fruit matures. eaches and tomatoes are fleshy fruits.

ower The reproductive organ of angiosperms nd gymnosperms

ond The leaf of a fern or a palm tree

uit In angiosperms, the mature ovary or roup of ovaries with or without other parts f the flower, containing the seeds

ngi Kingdom of single-celled and many-lled organisms that have nuclei and are nable to carry out photosynthesis

ametangium A cell or an organ within which ametes are formed

amete A haploid reproductive cell (n) that ust fuse with another gamete of the oppo-te sex to produce a diploid cell ($2n$) called zygote. From the zygote a new organism n develop.

ametophyte In the life cycle of a diplohap-nt plant, the haploid form (n) that pro-ices gametes through mitosis

astric juice A thin, watery digestive fluid creted by glands in the stomach

nes A unit of heredity located on a chro-osome

rmination The initial stages in the growth a seed

mnosperm A plant in which the seeds are ot enclosed within an ovary

noecium In flowers, the female reproduc-ve organ formed from one or more pistils

aploid Having a single rather than a dual t of chromosomes. Haploid organisms can produced as spores that develop through itosis and produce a gametophyte, or as ametes that must unite to develop.

aplont cycle A life cycle with a single hap-id generation (n), as in many algae and ngi. Meiosis occurs immediately after sex-al reproduction.

haustorium A branching filament typical of parasitic plants that penetrates the host and absorbs nutrients

hemiparasite A parasitic plant that can carry out photosynthesis

herbaceous Having soft, green stems

hermaphrodite An organism that has both male and female sexual organs, such as the stamens and pistils in flowers

heteromorphic alternation In diplohaplont cycles, the different haploid and diploid generations

heterospory The production of more than one type of spore

holdfast A rootlike structure that anchors a plant to something solid

indehiscent Dry fruit, such as that of maples and buttercups, that remains closed to protect the seed

indusium In ferns, a membrane that protects the sori

inflorescence A number of closely set flowers arranged according to a regular structure, which may appear to be a single large flower, such as the capitulum of a daisy

infructescence A compound fruit deriving from an inflorescence in which the single ovaries are fused together, as in figs and pineapples

isomorphic alternation A diplohaplont cycle that produces identical haploid and diploid generations having different numbers of chromosomes (n or $2n$)

isospore In ferns, the production of identical male and female spores

leaf Part of a plant that has chlorophyll and is capable of photosynthesis

legume A dry fruit that contains a number of seeds enclosed in pods

liverwort A bryophyte with a flat leaflike thallus or three rows of leaves growing on a stem anchored to the ground by rhizoids

macrospore In heterosporic plants, the female (haploid) spore producing a female gametophyte

macrosporophylls Fertile leaves that carry female gametangia

meiosis A process in which a cell with diploid chromosomes produces four cells with haploid chromosomes (gametes or spores), which are not identical to the original. The genetic material is mixed during cell division.

meiospore A spore that develops through meiosis. Gametophytes develop from meiospores.

meristem tissue Tissue within the adult plant that retains embryonic characteristics and is responsible for primary and secondary growth

microclimate The local climate of a habitat

microorganism A single-celled and very small, many-celled organism

microspore In heterosporic plants, the male (haploid) spore producing a male gametophyte

microsporophylls Fertile leaves that carry male gametangia

midrib The main vein or nerve of a leaf

mitosis Asexual cell reproduction in which the nucleus reproduces an exact copy of itself

monocarpic Flowering and bearing fruit only once. Monocarpic plants may be annuals, such as rice, or perennials, such as bamboo and agave.

monoecious Having the stamens and pistils develop in separate male and female flowers on the same plant

monosperm Fruit containing a single seed

multisperm Fruit having more than one seed

nectar A sweet liquid secreted by some plants

nucleus Large structure present in most cells. It controls all cell activities.

nutrients Chemical substances living things need for growth, energy, and repair

oomycete Algaelike fungi, such as water molds, white rusts, and downy mildews

organism Any particular form of life

ovary In the flowers of the angiosperms, the swollen base of the pistil in which the ovules develop. The mature ovary develops into a fruit.

ovule In seed plants, the structure containing the female gametophyte with the egg cell. When mature, the ovule becomes a seed.

parasitism Relationship in which an organism lives in or on another and causes harm to it

peduncle The stem that supports flowers and inflorescences of angiosperms

perennial A plant that lives for more than two years

perianth In flowers, the combined calyx and corolla

permeable Having openings that allow liquids or gases to pass through

petals In flowers, the elements forming the corolla. In most species, the petals are thought to be stamens that have lost their reproductive functions and have assumed the role of attracting pollinating insects or birds.

phaeophytes Brown algae

photosynthesis Process in which organisms having chlorophyll, such as plants, algae, and blue-green bacteria, use sunlight, carbon dioxide, and water to make food in the form of glucose and oxygen

pistil In flowers, the female fertile leaf modified to protect the ovule. It is made up of the ovary, the style, and the stigma—structures adapted to receive the pollen and facilitate fertilization.

pistillate A unisex female flower with a pistil but no stamen. It may be carried by separate sex (dioecious) or hermaphrodite (monoecious) plants. In the latter case, the same individual also carries male flowers.

pollen sac In the spermatophytes, the cavity in the anther containing the grains of pollen

pollination In the angiosperms, the transfer of the pollen from the male anther, where it was produced, to the female stigma

polycarpic Able to survive a number of annual cycles and bear fruit more than once

pome A fruit that has a leathery core with a seed that grows from the ovary or ovaries of a flower

postmaturation The sequence of processes whereby the seed germinates only when environmental conditions will allow the survival of the young plant

primary growth The growth resulting from the division and expansion of cells in the meristem

protonema In the bryophytes, a thin branching filament that occurs as a result of the germination of the spore. It grows buds that develop into gametophytes.

pteridophytes A group of plants made up of ferns, club mosses, and horsetails

receptacle In flowers, the tip of the peduncle on which all the elements of the flower itself are located

reproductive organs The organs within which the processes necessary for producing offspring occur

reproductive phase Part of the plant life cycle during which reproductive organs develop and sexual reproduction occurs

rhizoid A rootlike structure

rhizome An underground stem from which aerial branches that appear to be separate plants grow

rhodophytes Red algae

root The underground part of the plant that anchors it to the ground and supplies it with water and mineral salts

samara A simple dry fruit with a single seed that does not open when mature. It is carried by the wind on winglike membranes. Elms, maples, and ash have samaras.

secondary growth The increase in thickness of a stem or root

seed In the spermatophytes, the organ destine to reproduce the plant. The seed forms fro the maturation of the ovule.

seed dispersal Process of scattering seed often involving wind, water, or animals

seedling A young plant (sporophyte) deve oping from the embryo. In spermatophytes develops following the germination of th seed.

self-pollination Movement of pollen from stamen to a pistil of the same flower, or a di ferent flower on the same plant

self-sterile Being unable to fertilize the ovu by transferring pollen to the pistil of th same individual plant

sepal In flowers, the modified leaf that ac as the outer covering at the base of th flower. The sepals are usually green an capable of photosynthesis, although in som flowers the sepals look like petals.

sexual reproduction The union of two hap loid cells (*n*) that produce a zygote, or diploid cell (*2n*)

sori In ferns, a group of sporangia

sorosis The fleshy multiple fruit formed fro the infructescence of plants, such as the mu berry and the pineapple

spermatophytes Seed plants including th gymnosperms and the angiosperms

spermatozoa Male gamete

sporangia Hollow structures within whic spores are formed

spore A reproductive cell capable of deve oping into a many-celled adult organis without fusing with another cell; for exar ple, the gametophyte of a moss

sporocarp In ferns, a fold in the leaf that pr tects the sporangia

sporophyte A diploid organism *(2n)* form from the fusion of two gametes that pr duces spores through meiosis

stalk In mosses, the filamentary part of th sporophyte that supports the capsule

tamen In flowers, the male fertile leaf that ontains the structures that produce pollen

taminate In a flower, having only male eproductive organs

tigma In flowers, the swollen part of the pis-l that accepts and retains the pollen grain

tomata Tiny openings on the surface layer f a leaf that allow gases in and out

tyle In angiosperms, the long part of the pis-l that lies above the ovary

yconium The infructescence of the fig plant

symbiosis A relationship between two species that live in association with each other, providing benefits (food or protection) to both organisms

tepal In flowers, the name for the perianth in which petals and sepals cannot be distinguished

umbel An inflorescence in which the buds are flat-topped and the side stems are all attached at the tip of the stem. Carrots and dill are plants with umbels.

vascular Relating to the channel for the transport of a fluid

vegetative phase First part of the plant life cycle in which the plant develops all the organs needed for sexual maturity, such as roots, a stem, and leaves

xanthophytes Yellow-brown algae

zygote A diploid cell *(2n)* forming from the union of the male and female gametes. The zygote divides to create the sporophyte in the diploid and diplohaplont cycles, and immediately creates the meiospores in the haploid cycle.

Further Reading

Bates, Jeffrey W. *Seeds to Plants: Projects with Biology*. Watts, 1991
Capon, Brian. *Plant Survival: Adapting to a Hostile World*. Timber, 1994
Catherall, Ed. *Exploring Plants*. Raintree Steck-Vaughn, 1992
Cochrane, Jennifer. *Nature*. Watts, 1991
Facklam, Howard and Facklam, Margery. *Plants: Extinction or Survival?* Enslow, 1990
Garassino, Alessandro. *Plants: Origin and Evolution*. Raintree Steck-Vaughn, 1994
Greenway, Theresa. *Ferns*. Raintree Steck-Vaughn, 1992
———. *First Plants*. Raintree Steck-Vaughn, 1990
———. *Mosses and Liverworts*. Raintree Steck-Vaughn, 1992
Harlow, Rosie and Morgan, Gareth. *Trees and Leaves*. Watts, 1991
Landau, Elaine. *Endangered Plants*. Watts, 1992
Madgwick, Wendy. *Flowering Plants*. Raintree Steck-Vaughn, 1990
Peacock, Graham and Hudson, Terry. *Exploring Habitats*. Raintree Steck-Vaughn, 1992
Plants, Dorling Kindersley, 1992
Pope, Joyce. *Plant Partnerships*. Facts on File, 1991
Tant, Carl. *Seeds, etc. . .* .Biotech, 1992
Wiggers, Raymond. *Picture Guide to Tree Leaves*. Watts, 1992

INDEX

Picture Credits

The first number refers to the page. The number in parentheses refers to the illustration.

Photographs
Duilio Citi

U. BÄR VERLAG, Zürich (MAXIMILIEN BRUGGMANN): 50. ALBERTO CONTRI, Milan: 18 (b). RAFFAELE FATONE, Peschiera Borromeo, Milan: 35–36 (5,6), 37 (9). FOTORICERCA DONADONI, Bergamo: 11 (2,5). WALTER GRANERI, Milan: cover, 38 (all), 39 (all), 44–45 (1). EDITORIAL JACA BOOK, Milan (DUILIO CITI): 2–3, 11 (6), 19 (d), 20 (1), 21 (2), 23 (2,3), 24 (1,2), 25 (3), 26 (2), 28 (1,2), 29 (3,4), 30 (all), 31 (all), 36 (3,7), 37 (2, 10, 11), 40 (1), 41 (2), 44 (2), 45 (3,4,), 46 (1, 2, 3, 4), 48 (1), 49 (2), 51 (left); (GIORGIO DETTORI): 10; (RENATO MASSA): 11 (3, 4), 19 (c), 27 (3), 34, 36 (1, 4); (CARLO SCOTTI): 8. MIREILLE VAUTIER, Paris: 51 (right).
Color plates and drawings
EDITORIAL JACA BOOK, Milan (REMO BERSELLI): 41 (4); (MARIA ELENA GONANO):12–13, 26–27, 32–33, 42–43; (FABIO JACOMELLI and MAURIZIO GRADIN): 12–13 (1), 14–15 (1); (ROSALBA MORIGGIA & MARIA PIATTO): 13 (2), 14 (2), 16 (1, a, 2), 17 (3), 20 (1), 21 (4, 5, 6), 24 (1), 35, 36 (8), 40 (3); (LORENZO ORLANDI): cover; (GIULIA RE): 18 (4, 5), 19 (3), 41 (5).

Separate illustrations
pages 4–5: Green algae clinging to the rocks on the Ligurian coast of the Mediterranean Sea
page 8: The positive and negative images of a fossilized branch of *Annularia*, a giant horsetail of the genus *Calamites*. From the Carboniferous period, 280 to 345 million years ago, Marzon Creek, Illinois
page 50: A coniferous forest on Baranof Island in the Alexander Archipelago, Alaska, silhouetted against the late afternoon sun
page 51: Left: Autumn in a deciduous broadleaf forest in the Ligurian Apennines, Italy Right: A tropical forest halfway up the Marañon River, a tributary of the Amazon River